Communications
in Computer and Information Science 1988

Editorial Board Members

Joaquim Filipe ⓘ, *Polytechnic Institute of Setúbal, Setúbal, Portugal*
Ashish Ghosh ⓘ, *Indian Statistical Institute, Kolkata, India*
Lizhu Zhou, *Tsinghua University, Beijing, China*

Rationale

The CCIS series is devoted to the publication of proceedings of computer science conferences. Its aim is to efficiently disseminate original research results in informatics in printed and electronic form. While the focus is on publication of peer-reviewed full papers presenting mature work, inclusion of reviewed short papers reporting on work in progress is welcome, too. Besides globally relevant meetings with internationally representative program committees guaranteeing a strict peer-reviewing and paper selection process, conferences run by societies or of high regional or national relevance are also considered for publication.

Topics

The topical scope of CCIS spans the entire spectrum of informatics ranging from foundational topics in the theory of computing to information and communications science and technology and a broad variety of interdisciplinary application fields.

Information for Volume Editors and Authors

Publication in CCIS is free of charge. No royalties are paid, however, we offer registered conference participants temporary free access to the online version of the conference proceedings on SpringerLink (http://link.springer.com) by means of an http referrer from the conference website and/or a number of complimentary printed copies, as specified in the official acceptance email of the event.

CCIS proceedings can be published in time for distribution at conferences or as post-proceedings, and delivered in the form of printed books and/or electronically as USBs and/or e-content licenses for accessing proceedings at SpringerLink. Furthermore, CCIS proceedings are included in the CCIS electronic book series hosted in the SpringerLink digital library at http://link.springer.com/bookseries/7899. Conferences publishing in CCIS are allowed to use Online Conference Service (OCS) for managing the whole proceedings lifecycle (from submission and reviewing to preparing for publication) free of charge.

Publication process

The language of publication is exclusively English. Authors publishing in CCIS have to sign the Springer CCIS copyright transfer form, however, they are free to use their material published in CCIS for substantially changed, more elaborate subsequent publications elsewhere. For the preparation of the camera-ready papers/files, authors have to strictly adhere to the Springer CCIS Authors' Instructions and are strongly encouraged to use the CCIS LaTeX style files or templates.

Abstracting/Indexing

CCIS is abstracted/indexed in DBLP, Google Scholar, EI-Compendex, Mathematical Reviews, SCImago, Scopus. CCIS volumes are also submitted for the inclusion in ISI Proceedings.

How to start

To start the evaluation of your proposal for inclusion in the CCIS series, please send an e-mail to ccis@springer.com.

Xingwei Wang · Mingwei Xu · Xiaoqiu Shi ·
Fan Wu
Editors

Frontiers of Networking Technologies

First China Conference on Networking, CCF ChinaNet 2023
Wenzhou, China, September 8–10, 2023
Proceedings

Editors
Xingwei Wang
Northeastern University
Shenyang, China

Mingwei Xu
Tsinghua University
Beijing, China

Xiaoqiu Shi
Wenzhou University
Wenzhou, China

Fan Wu
Shanghai Jiao Tong University
Shanghai, China

ISSN 1865-0929 ISSN 1865-0937 (electronic)
Communications in Computer and Information Science
ISBN 978-981-97-3889-2 ISBN 978-981-97-3890-8 (eBook)
https://doi.org/10.1007/978-981-97-3890-8

© The Editor(s) (if applicable) and The Author(s), under exclusive license to Springer Nature Singapore Pte Ltd. 2024

This work is subject to copyright. All rights are solely and exclusively licensed by the Publisher, whether the whole or part of the material is concerned, specifically the rights of translation, reprinting, reuse of illustrations, recitation, broadcasting, reproduction on microfilms or in any other physical way, and transmission or information storage and retrieval, electronic adaptation, computer software, or by similar or dissimilar methodology now known or hereafter developed.
The use of general descriptive names, registered names, trademarks, service marks, etc. in this publication does not imply, even in the absence of a specific statement, that such names are exempt from the relevant protective laws and regulations and therefore free for general use.
The publisher, the authors and the editors are safe to assume that the advice and information in this book are believed to be true and accurate at the date of publication. Neither the publisher nor the authors or the editors give a warranty, expressed or implied, with respect to the material contained herein or for any errors or omissions that may have been made. The publisher remains neutral with regard to jurisdictional claims in published maps and institutional affiliations.

This Springer imprint is published by the registered company Springer Nature Singapore Pte Ltd.
The registered company address is: 152 Beach Road, #21-01/04 Gateway East, Singapore 189721, Singapore

If disposing of this product, please recycle the paper.

Preface

Hosted by China Computer Federation (CCF), the 2023 China Conference on Networking (CCF ChinaNet 2023) was held in Wenzhou on September 8–10, 2023. It was organized by CCF Technical Committee of Internet (TCI), CCF Technical Committee of Network and Data Communications (TCCOMM), and Wenzhou University and supported by Wenzhou Municipal People's Government. The theme of the conference was "Integrating Industry, Academia, Research, and Innovation, Serving the Networking Powerhouse".

CCF ChinaNet was upgraded with the combination of CCF Internet Conference of China (ICoC) and CCF Network and Data Communication Conference (NDCC), which are widely known in the area of computer science and engineering. The conference invited a number of top scholars, such as CAS/CAE Members, ACM/IEEE Fellows, and Changjiang/Outsanding-Youth scholars, to make high-level academic reports. At the same time, experts from well-known universities, research institutions, and enterprises were invited for academic and technical exchanges.

This volume contains 13 high-quality papers from CCF ChinaNet 2023, as selected from 61 submissions after a single-blind peer review process. Each paper was reviewed by 2–3 Program Committee members. We hope this volume will be beneficial for readers from both academia and industry, who may understand and tackle the challenges in an efficient manner and adopt appropriate solutions in the related fields.

September 2023

Xingwei Wang
Mingwei Xu
Xiaoqiu Shi
Fan Wu

Organization

General Chairs

Jinshu Su — NUDT, China
Keqiu Li — Tianjin University, China
Min Zhao — Wenzhou University, China
Huadong Ma — BUPT, China

Program Committee Chairs

Xingwei Wang — Northeastern University, China
Mingwei Xu — Tsinghua University, China
Xiaoqiu Shi — Wenzhou University, China
Fan Wu — Shanghai Jiao Tong University, China

Program Committee

Ziling Wei — NUDT, China
Ming Tao — Dongguan University of Technology, China
Xiaoling Tao — Guilin University of Electronic Technology, China
Honglong Chen — China University of Petroleum, China
Zhaomin Chen — Wenzhou University, China
Li Chen — Zhongguancun Laboratory, China
Yige Chen — Wenzhou University, China
Qingyong Deng — Guangxi Normal University, China
Chuan Zhao — University of Jinan, China
Gongming Zhao — University of Science and Technology of China, China
Yi Zhao — Tsinghua University, China
Shi Dong — Zhoukou Normal University, China
Yisu Ge — Wenzhou University, China
Lailong Luo — NUDT, China
Junling Shi — Shenyang Aerospace University, China
Shuai Wang — Zhongguancun Laboratory, China
Xianlong Jiao — Chongqing University, China
Zhiyong Pan — Wenzhou University, China

Shuo Wang	BUPT, China
Dan Tang	Hunan University, China
Lin Liu	NUDT, China
Fuliang Li	Northeastern University, China
Rongfei Zeng	Northeastern University, China
Dapeng Qu	Liaoning University, China
Bo Yi	Northeastern University, China
Yingying Xu	Wenzhou University, China
Wei Peng	NUDT, China
Shuzhuang Zhang	BUPT, China
Yuchao Zhang	BUPT, China
Shenglin Zhang	Nankai University, China
Shiwen Zhang	Hunan University of Science and Technology, China
Qingyuan Gong	Fudan University, China
Yongqian Sun	Nankai University, China
Zhenzhou Tang	Wenzhou University, China
Huan Zhou	NUDT, China
Qiang He	Northeastern University, China

Contents

Remote Sensing GEO Relay Transmission Scheme Based on Mission Awareness Under Dynamic Topology 1
    Jing Chen, Xiaoqiang Di, Hui Qi, Jinqing Li, and Ligang Cong

Review of Image Encryption Based on Bibliometrics and Visualized Analysis 17
    Zhaoyang Liu and Ru Xue

DDPG-FL: A Reinforcement Learning Approach for Data Balancing in Federated Learning 33
    Bei Ouyang, Jingyi Li, and Xu Chen

Adaptive Recovery with Reinforcement Learning in Cloud-of-Clouds Storage Systems 48
    Jiajie Shen, Bochun Wu, Wang Xiang, Zeyu Zhao, and Kai Zhang

DAG: A Lightweight and Real-Time Edge Defense Model for IoT DDoS Attacks 61
    Yanhua Liu, Cong Chen, Qiu Zhang, Fanhao Zeng, and Ximeng Liu

A Data Publishing Method for Trajectory Privacy Classification Based on Differential Privacy 74
    Qian He, Bingjie Liao, Peng Liu, and Qinghe Dong

Towards Anomaly Traffic Detection with Causal Interpretability Methods 84
    Zengri Zeng, Baokang Zhao, Xuhui Liu, and Xiaoheng Deng

Multi-class Intrusion Detection System in SDN Based on Hybrid LSTM Model 99
    Jue Chen and Meng Cui

Design and Implementation of Computing Based Service Chain Orchestration Framework 112
    Dongsheng Qian, Yusheng Lv, Kuo Guo, Shang Liu, Xu Huang, Chenxi Liao, Jingjing Liu, Xiaolong Liu, Kai Chen, and Jia Chen

Towards Smart Stream Scheduler for Multipath QUIC in Heterogeneous Networks 128
    Anyi Li, Xiangbin Liang, Tianshu Wang, and Baokang Zhao

VotePipe: Efficient Heavy Hitter Detection in Programmable Data Plane 146
  Danqi Li, Ningbo Tian, Kun Qiu, Harry Chang, Xiahui Yu, and Jin Zhao

SDN Based Network Path Planning Optimization for Printing Cloud
Service 167
  Jiajun Peng, Qian He, Qi Pan, and Yanbo Liu

A Survey on Security Issues of SDN Controllers 182
  *Rui Wang, Youhuizi Li, Meiting Xue, Baokang Zhao, Yuyu Yin,
  and Yangyang Li*

Author Index 207

Remote Sensing GEO Relay Transmission Scheme Based on Mission Awareness Under Dynamic Topology

Jing Chen[1,2,3], Xiaoqiang Di[3(✉)], Hui Qi[3], Jinqing Li[3], and Ligang Cong[3]

[1] School of Computer Science and Technology, Changchun University of Science and Technology, Changchun 130022, China
[2] Changchun Sci-Tech University, Changchun 130600, China
[3] Jilin Key Laboratory of Network and Information Security, Changchun University of Science and Technology, Changchun 130022, China
{dixiaoqiang,qihui,lijinqing}@cust.edu.cn

Abstract. With low-orbit (LEO) remote sensing being widely used in emergency rescue, atmospheric observation, and military reconnaissance, the types of users and their requirements are increasing. Due to the fact that the communication time between LEO remote sensing satellites and ground stations is only 10–15 minutes each time, which cannot meet the real-time service needs of a large number of users, using geostationary orbit (GEO) relay satellites to relay the transmission has become an effective solution. In practical applications, the payoffs of transmitting tasks of the same size but different levels via relay satellites are different. When multiple LEOs apply for data forwarding services, how to provide the best forwarding rate for tasks of different priorities will directly affect the efficiency of task completion and the revenue of GEOs. Most of the current methods do not consider the impact of topology changes on the policy, and cannot be well applied to satellite networks. To solve the above problems, this paper proposes a GEO relay transmission scheme (GRT-TADT) based on the task awareness under dynamic topology, and a priority-based GEO resource allocation algorithm, which can achieve the optimal allocation of the GEO rate under each time slot according to the task priority to maximize the GEO gain. The experimental results show the effectiveness of the GRT-TADT scheme in terms of the number of tasks completed, GEO resource allocation, and revenue. Compared to the GAA-FARR strategy, the GRT-TADT scheme is 48.2% higher than the GAA-FARR strategy in terms of GEO revenue when transmitting the same amount of data.

Keywords: Satellite networks · Task awareness · Relay transmission · GEO resource allocation · Remote sensing data

---

This work was funded by the National Natural Science Foundation of China (U21A20451),the China University Industry-Academia-Research Innovation Fund (2021FNA01003).

1 Introduction

With the development of earth satellite observation systems, the extensive use of global coverage satellites, high-precision reconnaissance satellites, high-resolution satellite systems, etc., the volume of downstream data has exploded while the volume of demand-based remote sensing data transmission tasks has also increased dramatically [12]. As LEO remote sensing satellites operate periodically in their orbits, the communication between the satellites and the ground stations is also periodic, only communicating within each other's visibility time windows (VTW), the satellites will use different modes of operation to transmit back within the VTW according to the user's needs [1], but the VTW is usually around 10–15 minutes [13], and the amount of data that can be transmitted down is very limited. Therefore, how to send data from LEO remote sensing satellites to the ground promptly has become a key issue in remote sensing applications [19]. In recent years, GEO satellite data forwarding has been proposed as an effective solution [14]. Xu et al. [18]used GEO satellites to forward remote sensing data in real-time, which improved in terms of throughput and transmission delay(GAA-FARR). Wang et al. [15] proposed a single GEO to multi-LEO content caching scheme from the perspective of content popularity to alleviate network congestion and improve system throughput. Deng C et al. [5] used a relay network consisting of multiple GEOs to serve LEO networks, but they mainly focused on problems such as routing in the network. Wang L et al. [16] proposed an efficient resource allocation scheme for data relay satellite systems, designing penalty and forgiveness strategies based on user behavior, and reducing resource conflicts in a cooperative manner to maximize user benefits. Du J et al. [7] proposed a GEO relay bandwidth resource allocation strategy aimed at maximizing network throughput. The above-mentioned literature designs GEO resource allocation schemes from different perspectives. With the development of remote sensing applications, there are more and more user types, and different user types have differentiated characteristics (e.g. military intelligence, meteorological monitoring, urban planning, etc.) [10], which will lead to the differentiated value of remote sensing image data to meet the demand. When a single GEO provides forwarding services for multiple LEOs, the efficient use of GEO resources for different user requirements has a significant impact on the efficiency of task completion. At the same time, the value brought by transmitting different mission data is also different, which will directly affect the benefit of GEO [6]. In addition, due to the high-speed movement of LEO remote sensing satellite nodes, the LEO nodes involved in applying for forwarding services vary from time slot to time slot, i.e. the change in topology is also crucial to the strategy [8].

In order to solve the problem of fast return transmission of remote sensing data demanded by users in satellite networks, this paper proposes a real-time remote sensing GEO relay transmission scheme based on task awareness under dynamic topology. In this paper, we consider the time-varying characteristics of the inter-layer topology, the size and priority of tasks, and design a user-demand-based GEO resource allocation strategy using a network utility maximization framework [11], with the objective of maximizing the revenue of GEO under each time slot.

The contributions of this paper are summarized as follows:

- To address the problem of real-time remote sensing data transmission, design a GEO relay transmission scheme for real-time remote sensing based on task awareness(GRT-TADT), which provides different forwarding rates according to task priority and aims to maximize GEO benefits.
- To address the problem of real-time allocation of GEO resources, a resource-on-demand allocation algorithm under dynamic topology is designed according to user requirements using a network utility maximization framework to achieve high demand corresponding to high forwarding rates.
- Simulation experiments have confirmed the effectiveness of GRT-TADT in terms of the number of tasks completed, allocation of GEO resources, and benefits. Compared with the GAA-FARR strategy, GRT-TADT is 48.2% higher than the GAA-FARR strategy in terms of GEO benefits when transmitting the same amount of data.

The rest of the paper is organised as follows: Section 2 presents the network model. Section 3 presents the transmission scheme and algorithms. Section 4 does the performance evaluation and analyses the results. Section 5 concludes the paper.

2 Network Model

As shown in Fig. 1, consider a three-layer network architecture: 3 GEOs as the forwarding layer, 1 LEO remote sensing satellite constellation as the mission data source layer, and the ground layer including the ground control center and multiple users. Multiple LEO satellites can request forwarding services from one GEO at the same time. The LEO layer is based on the NDN architecture, and users can request data based on their interests(details in 3.2 Task description). When an LEO remote sensing satellite with task data passes over a user, the LEO remote sensing satellite transmits the data directly to the user, otherwise, the SDN controller assigns a suitable GEO relay satellite to it and then forwards the data to the user via the GEO relay satellite(i.e. using the LEO-GEO-User link). Assuming an ideal space communication environment, GEO can forward data while receiving it, regardless of signal attenuation. Denote GEO satellites by $G_n = \{G_1, G_2, G_3\}$, and the maximum transmission rate of G_n is Q_n. $S_m = \{S_1, S_2, \cdots, S_{M-1}, S_M\}$, denotes the set of LEO satellites and $m = 1, 2, \cdots, M$ denotes the number of LEO satellites. $U_u = \{U_1, U_2, \cdots\}$ denotes the set of users, $Task_u = (D_u, T_u, L_u)$ denotes the task applied by each user, D_u, T_u, L_u denotes the size, final completion time and level of the task respectively. S_m has task data $Task_u$ requested by U_u. In this paper, we assume that the amount of data forwarded by S_m to G_n per request is in task units.

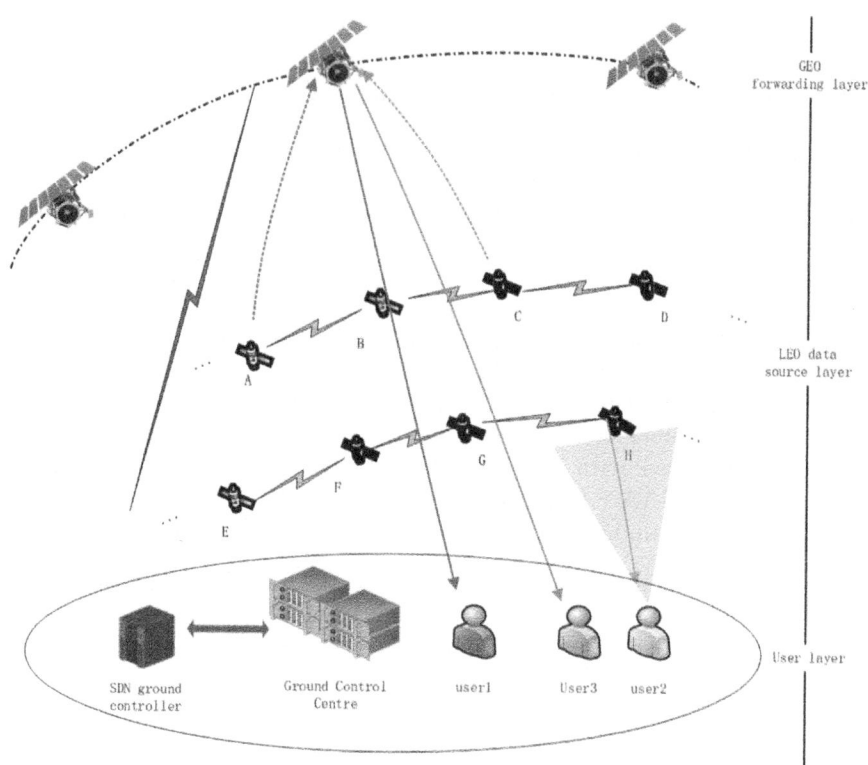

Fig. 1. Network architecture diagram

3 Remote Sensing GEO Relay Transmission Scheme Based on Mission Awareness Under Dynamic Topology—GRT-TADT

This section proposes a real-time remote sensing GEO relay forwarding scheme based on task awareness, which comprehensively considers the topology change, task size, and priority, and realizes the demand-based forwarding rate allocation. The modeling and solving process of the scheme is described in detail.

3.1 Link Selection

In this paper, NDN is combined with SDN. In a 6G direct link satellite network environment, the user sends interest to the remote sensing satellite network under the NDN architecture, which includes the level and completion time of the task.

In this paper, the visible relationships between the three layers of entities are defined as follows:

$$Z_{SU} = \begin{bmatrix} Z_{S_1U_1} & Z_{S_2U_1} & \cdots & Z_{S_mU_1} \\ Z_{S_1U_2} & Z_{S_2U_2} & \cdots & Z_{S_mU_2} \\ \vdots & \vdots & \ddots & \vdots \\ Z_{S_1U_k} & Z_{S_2U_k} & \cdots & Z_{S_mU_k} \end{bmatrix} \qquad (1)$$

$$Z_{GS} = \begin{bmatrix} Z_{S_1G_1} & Z_{S_2G_1} & \cdots & Z_{S_mG_1} \\ Z_{S_1G_2} & Z_{S_2G_2} & \cdots & Z_{S_mG_2} \\ Z_{S_1G_3} & Z_{S_2G_3} & \cdots & Z_{S_mG_3} \end{bmatrix} \qquad (2)$$

$$Z_{GU} = \begin{bmatrix} Z_{U_1G_1} & Z_{U_2G_1} & \cdots & Z_{U_kG_1} \\ Z_{U_1G_2} & Z_{U_2G_2} & \cdots & Z_{U_kG_2} \\ Z_{U_1G_3} & Z_{U_2G_3} & \cdots & Z_{U_kG_3} \end{bmatrix} \qquad (3)$$

$$Z_{SGU} = Z_{GS} \cap Z_{GU} \qquad (4)$$

where $Z = 0, 1$ denotes the visible relationship between two entities, $Z = 1$ means visible and $Z = 0$ means invisible. Eq. (1) represents the visible relationship between the LEO and the user, Eq. (2) represents the visible relationship between the LEO and the GEO, Eq. (3) represents the visible relationship between the user and the GEO, and Eq. (4) represents the task using the GEO for relay forwarding. During the path selection process, if $Z_{SU} = 1$, $U_k S_{me} \geq T_u$ and $D_u \leq B_{SG} * T_{SU}$, the LEO-U link is used for transmission, otherwise the SDN controller selects a suitable G_n for it based on whether Z_{SGU} is 1 and forwards the transmission according to the level, where $U_k S_{me}$ indicates the end time of the user and LEO visible time, T_u indicates the latest time when the current task is completed, D_u indicates the task size and $B_{SG} * T_{SU}$ indicates the throughput of the LEO-U link, which is the product of the bandwidth and the effective time of the link.

3.2 Task Description

Remote sensing data has space-time attributes, users can submit their requirements to the ground control center based on this information, the control center will send the task to the remote sensing satellite network under the NDN architecture for data retrieval, with reference to the naming mechanism under the NDN architecture, the interest package sent by the user is as follows [3]:

$$interest : /S_{\#}/type/loc/time/L_u \qquad (5)$$

where $S_{\#}$ denotes the LEO remote sensing satellite, which is initially identified by # as the user does not know the satellite number where the mission data is located, and can be replaced by the corresponding satellite number when the packet is returned after retrieving the data source node, and $/type/loc/time/L_u$ refers to the specifics of the user's requirements — including type, spatial attributes, temporal attributes, and corresponding levels. The "L_u"

part can be defined by the user in accordance with their own requirements and market rules.

So the corresponding packets are named as follows:

$$data : /S_m/type/loc/time/D_u/L_u \tag{6}$$

where D_u indicates the task size, which can be calculated from the source information of each remote sensing image data.

4 GEO Resource Allocation

When the network selects LEO-GEO-U links for task transmission, how to allocate the link bandwidth of G_n among multiple tasks is addressed in this paper based on the utility maximization model, which aims to maximize the G_n gain. The size of the task $Task_U$ that exists on S_m that meets the requirements of user U_u is denoted by D_u, p_{smu} denotes the forwarding rate of S_m to G_n for $Task_U$ requests. The cost function for S_m is denoted by $F_u(p_{smu})$ and represents the spend, which is positively related to the task level, i.e. if the task size is the same, the higher the level, the higher the spend. q_{smu} denotes the forwarding rate offered by G_n to S_m. The utility function of G_n is the difference between the benefit and the cost, where the benefit is the spend of S_m and the total cost function is defined as $V(\sum_{u,m=1}^{U,M} q_{smu})$, which is the consumption of the forwarding. The objective is to maximize the utility function of G_n with the following objective function:

$$maxmize : \sum_{u,m=1}^{U,M} F_u(p_{smu}) - V\left(\sum_{u,m=1}^{U,M} q_{smu}\right) \tag{7}$$

$$s.t. \sum_{u,m=1}^{U,M} q_{smu} \leq Q_n \tag{8}$$

$$p_{smu} \leq q_{smu} \tag{9}$$

$$p_{smu} \geq 0, q_{smu} \geq 0 \tag{10}$$

For the above objective function, the SDN controller has control over the global information. In each time slot, S_m submits D_u and L_u information to the SDN controller, while G_n submits available forwarding rate q_{smu} information to the SDN controller. Based on q_{smu}, D_u, and L_u, the SDN controller determines the actual forwarding rate p_{smu} and cost for each S_m.

4.1 Function Definitions

Referring to the rate allocation mechanism proposed by Q. Xu et al. [17], $F_u(p_{smu})$ and $V(\sum_{u,m=1}^{U,M} q_{smu})$ are defined as follows, respectively:

$$F_u(p_{smu}) = Z_{SGU} * D_u * L_u * \log p_{smu} \qquad (11)$$

$$V(\sum_{u,m=1}^{U,M} q_{smu}) = C_n * \sum_{u,m=1}^{U,M} \frac{q_{smu}^2}{2} \qquad (12)$$

where C_n denotes the unit cost of G_n and q_{smu} is related to L_u and Q_n, so the objective function is converted as follows:

$$maxmize : Z_{SGU} * D_u * L_u * \log p_{smu} - C_n * \sum_{u,m=1}^{U,M} \frac{q_{smu}^2}{2} \qquad (13)$$

$$s.t. \sum_{u,m=1}^{U,M} q_{smu} \leq Q_n \qquad (14)$$

$$p_{smu} \leq q_{smu} \qquad (15)$$

$$p_{smu} \geq 0, q_{smu} \geq 0 \qquad (16)$$

4.2 Objective Function Solution

Solving for the above objective function can be done using the Lagrange multiplier method and the KKT condition, the Lagrangian function of Eq. (13) is as follows:

$$\begin{aligned} L(p_{smu}, q_{smu}, \lambda_u, \mu_u) = & Z_{SGU} * D_u * L_u * \log p_{smu} \\ & - C_n * \sum_{u,m=1}^{U,M} \frac{q_{smu}^2}{2} \\ & - \lambda_u (\sum_{u,m=1}^{U,M} q_{smu} - Q_n) \\ & - \sum_{u,m=1}^{U,M} \mu_u (p_{smu} - q_{smu}) \end{aligned}$$

Here λ_u, μ_u are the Lagrange multipliers for each of the two constraints, and according to the Karush-Kuhn-Tucker (KKT) [2] optimality conditions the following equation holds:

(A1) $\frac{\partial L}{\partial p_{smu}} = \frac{Z_{SGU}*D_u*L_u}{p_{smu}} - \mu_u = 0$.

(A2) $\frac{\partial L}{\partial q_{smu}} = -C_n * q_{smu} - \lambda_u + \mu_u = 0$.

(A3) $\lambda_u(\sum_{u,m=1}^{U,M} q_{smu} - Q_n) = 0$.

(A4) $\mu_u(p_{smu} - q_{smu}) = 0$.

Equation (11) increases strictly with p_{smu} and is related to the task level, in the best case, there is $p_{smu} = q_{smu}$, corresponding to the expression 4.2 = 4.2, i.e:

$$\frac{Z_{SGU}*D_u*L_u}{p_{smu}} - \mu_u = -C_n * q_{smu} - \lambda_u + \mu_u \tag{17}$$

Solve for μ_u:

$$\mu_u = \frac{Z_{SGU}*D_u*L_u + C_n * q_{smu} * p_{smu} + \lambda_u * p_{smu}}{2 * p_{smu}} \tag{18}$$

μ_u is the transaction price and λ_u is the forwarding rate at the time of the deal. According to the rate allocation rules of Kelly F P et al. [9], the forwarding rate provided for S_m is:

$$\lambda_u = \frac{p_{smu} * L_u}{\mu_u} \tag{19}$$

Substitute Eq. (18) into Eq. (19) to solve for λ_u:

$$\lambda_u = \frac{-b + \sqrt{b^2 + 8 * p_{smu}^3 * L_u}}{2 * p_{smu}} \tag{20}$$

where $b = Z_{SGU} * D_u * L_u + C_n * q_{smu} * p_{smu}$.

The quadratic derivation of 4.2 and 4.2 shows that both the G_n benefit function and the cost function are strictly concave functions, so there is a maximum value that ensures that the utility of G_n is maximized.

4.3 GEO Benefits

Based on the above description, the benefits of G_n are as follows:

$$E_n(p_{smu}, p_{gmu}) = -C_n * \sum_{u,m=1}^{U,M} \frac{q_{smu}^2}{2} + \sum_{u,m=1}^{U,M} \mu_u \tag{21}$$

G_n finds the optimal transmission rate vector λ_u for each task by solving the above gain maximization problem: G_n solves the maximization gain problem by maximizing the result of Eq. (21), and thus obtains the optimal transmission rate vector λ_u for each task:

$$G_n - MAX : max\ E_n(p_{smu}, p_{gmu}) \tag{22}$$

$$s.t. q_{smu} \geq 0 \tag{23}$$

To obtain a unique optimal solution to Eq. (22), the following conditions need to be satisfied, the first-order derivative of Eq. (21) has:

$$\frac{\partial V(\sum_{u,m=1}^{U,M} p_{gmu})}{\partial p_{gmu}} = -C_n * q_{smu} - \lambda_u + \mu_u \quad (24)$$

Continue with the second-order derivative of Eq. (21):

$$\frac{\partial V^2(\sum_{u,m=1}^{U,M} p_{gmu})}{\partial p_{gmu}^2} = -C_n \quad (25)$$

Because the second order derivative is negative ($-C_n > 0$), it is clear that Eq. (21) is a strictly concave function, so there is a maximum value. Based on the above analysis, it can be ensured that G_n maximizes utility (Eq. (13)).

4.4 GEO Resource Allocation Algorithm

The SDN controller under each time slot allocates GEO resources based on the transaction price in Eq. (18) and the rate allocation rules in Eq. (20) to maximize the benefits of GEO. See Algorithm 1 for details.

Algorithm 1. GEO resource allocation algorithm

Require:
1: At each time slot, S_m and G_n submit messages p_{smu}, p_{gmu}, update the information submitted by each participant within each time slot.
2: Calculation: The SDN controller calculates μ_u, λ_u for each time slot based on Eq. (18) and Eq. (20). SDN publishes μ_u, λ_u.
3: G_n forwards S_m's task data based on λ_u and S_m pays at μ_u.
4: G_n calculates the benefit based on Eq. (21).

Ensure:
5: μ_u, λ_u, G_n 's benefits;

---

5 Experimental Results and Analysis

In the simulation scenario there are 3 GEO (theoretically, 3 GEOs can provide global coverage) and 1 LEO remote sensing constellation under NDN architecture (there are 6 orbits, and 6 satellites are evenly deployed in each orbit), multiple ground users in the cities of Kashgar, Miyun, and Sanya where the three ground stations are located, the users send task requests to the control center, and the control center makes the transmission plan for each task based on the SDN controller's determination. This paper focuses on the impact of tasks with the same size but different priorities on the GEO gains, so the task transmission when the LEO crosses the top user area is not considered. The data forwarding rate requested by the LEO to the GEO is λ_u, and the data transmission rate from the GEO satellite to the ground is 450 Mbps. Each GEO can communicate with more than one LEO at the same time. The mission priorities range from 1 to 6, with 6 being the highest priority (6 levels are set because the LEO constellation has 6 orbits in the simulated environment).

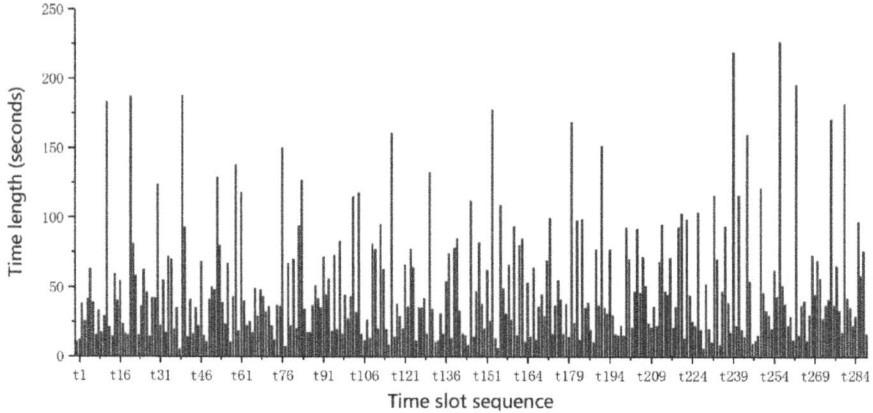

Fig. 2. Time slot division results

5.1 Dynamic Topology Analysis

Using STK to simulate the run data for 4 h (from 04h00 to 08h00), the data were processed using MATLAB, referring to our previous method in the literature [4], and taking G_1 as an example, the discretized time slots were obtained as shown in Fig. 2. The time slots ranged from 10 to 227 s in length, 4 h divided into 288 time slots.

Extracting the visible relationship between 5 o'clock and 6 o'clock (time slot t_{71} to t_{170}), the statistical number of visible time slots for S_m and G_1 is shown in Fig. 3. Taking s111 as an example, in these 100 time slots, 24 time slots are directly connected to the user, 51-time slots are linked to G1 and 25 time slots are linked to G2. If the data required by the user is on s111, the data can be transmitted to the user all the way through these 100 time slots.

5.2 Analysis of GEO Resource Allocation

Based on the above visible relations, the allocation of GEO resources was analyzed for 36 time slots (t71-t106), the time slot lengths (in seconds) are shown in Table 1. In this paper, we analyze the performance of the scheme in terms of the same task size (50M per task size) but with different priorities, and analyze five metrics: the rate allocation of GEO, the number of tasks completed under

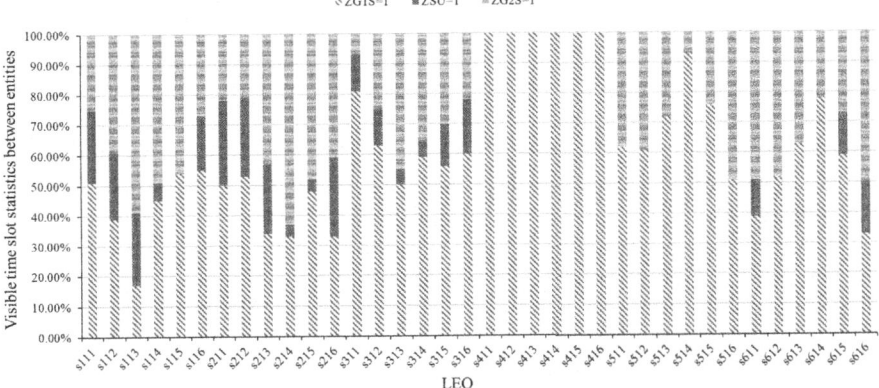

Fig. 3. time slot statistics for S_m with G_n and U_u

each time slot, the cost of LEO, the resource utilization of GEO and the benefit of GEO, as well as to evaluate the performance of the scheme and to compare it with GAA-FARR [18] in terms of benefit.

Table 1. Time slot and corresponding length (seconds)

time slot	t71	t72	t73	t74	t75	t76	t77	t78	t79	t80	t81	t82
length	36	22	12	37	36	157	67	22	70	20	94	127
time slot	t83	t84	t85	t86	t87	t88	t89	t90	t91	t92	t92	t94
length	34	17	17	37	51	42	35	72	44	56	18	73
time slot	t95	t96	t97	t98	t99	t100	t101	t102	t103	t104	t105	t106
length	19	83	16	44	27	43	115	32	118	16	11	26

The priority of the tasks in each LEO at different time slots is shown in Fig. 4. The highest priority is 6. The no-level cell indicates that the corresponding LEO satellite at the current time slot does not request a forwarding rate from the GEO satellite. Subsequent data analysis is done at the current task level setting.

As shown in Fig. 5, when the tasks are of the same size, the rates requested by tasks of different priority levels under different time slots are different. Taking time slot t71 as an example, the LEO satellites involved in the request for forwarding are S112, S212, S311, S312, S314, S315, S316, S411-S146, S511, S512, S515, S516, S612, S615 and S616, the higher the priority level the faster the requested forwarding rate.

LEO / time slots	s111	s112	s113	s114	s115	s116	s211	s212	s213	s214	s215	s216	s311	s312	s313	s314	s315	s316	s411	s412	s413	s414	s415	s416	s511	s512	s513	s514	s515	s516	s611	s612	s613	s614	s615	s616
t71		5						3					4	5		1	5	5	4	3	4	5	5	2	5	3			5	2		5				2
t72		1						3					5	6		1	3	3	5	2	4	5	2	4	1	2			3	5		4			3	6
t73		4						4					2	6		5	6		1	1	6	3	2		2	4			3	6		3			2	2
t74		5						4					3	5		5	3		5	2	2	4	4	2	2	4			3	5		3				5
t75		5						4		2	5		6	5		5	4		4	5	5	3	2	5	5	4			3	5		1			4	1
t76		2						4		4	6		5	1	5	5	1		4	3	6	3	1	1	4	4			6	3		3			3	3
t77	2	6		4				3		4	8		5	2	4	2	2		2	5	3	2	5	6	5	5			6	3		5			5	4
t78	5	1		5				5		5	6		4	5	6	3	2		4	3	2	4	3	6	4			5	5	2		1			5	3
t79	3	2		4				2		4	4		3	5	1	5	2		5	2	4	2	4	1	4			1	1	5		1			6	3
t81	4	6		3				4		2	2		5	5	4	3	4		2	4	3	6	2	5	4			5		3					3	3
t81	3	6		5				5		5	2		5	4	3	6	2		3	5	6	1	4	5	1			6		5					6	4
t82	4	4		2						5	1		4	2	1	5	5		6	5	4	2	5	2	2			5	6	3					5	5
t83	6	4		5				6		5	5		5	3	5	4	5		5	2	6	3	2	5	2			5	5	5	5				4	5
t84	4			4				1		5	5		5		5	1	5		2	4	6	2	5	2	4			4	2	4	6				2	5
t85	4	6		4				3		6	5		3		2	5	3		4	5	5	4	6	4	2			4	4	4	2			4	4	3
t86	6	4		3				2		6	5		6		5	2	2		5	6	2	4	3	3	3			5	5	3	4			2	4	5
t87	6	5		3				5		5			2		5	4	2		4	5	3	3	5	3	3			4	3	5	5			2	4	
t88	4	3						3		4			6		4	6	5		5	5	5	3	5	5	2			6	3	6	6			4		
t89	4	4		5						4			1		5	6			5	5	4	5	3	2	5			3	5	5	5			5	2	
t91	6	6								1			4		5	4			2	2	2	5	1	4	4			3	4	4	6			4		
t91	3	3					5			4			5		2	4			4	4	5	4	6	5	5			3	6	5	2			4	6	
t92	1						4			3			4		4	5			3	1	6	4	5	2	2				3	3	5			1	4	
t93	4						6			3			5		5	4			6	6		4	6	4	1			5	3	3	4			6	5	
t94	3						4		5	2			1		5	5			4	6	6	4	5	5	4			6	5	4	5			3	3	
t95	6		3				4		5				2		3	5		4	5	4	5	5	4	4	4			1	6	4	5			1	2	
t96	2		5				3		4	5			5	5	3	2		4	6	5	5	5	5	1				3	5	4	5			4	5	
t97	5						4		5	2			5	3	5	2		5	5	6	2	5	6	5	2			4	5	4	1			2	5	
t98	2		5			3	6			5			5	2	5	4		4	4	5	6	2	5	3	5			3	4	3	5			2	2	
t99	5		5			4	5		4	3			5	3	4	5		5	3	4	4	5	2	2	2		3	5	4	5	5			2	4	
t111	3		5			3	1		7	6			3	1	6	2		3	6	6	5	5	3	6	5		1	4	6	3	5			4	4	
t111	4		5			3	3		3	4			6	2	5	3		6	3	2	6	2	5	3			6	3	2	2				5	2	
t112	5		5			4	4		4	5			4	5	5	5		4	2	1	6		1	2			5	4	2					4		
t113	5		4			4	5		5	6			2	6	4	2		2	3	1	2	3	5	5			5	3	3	4				3	2	
t114	2		4			5	1		3	3				4	3	4		3	3	4	3	6	5	6			3	4	4	4				4	3	3
t115	5		3			3	5		5	6				4	5	4		3	2	5	2	3	5	4			5	5	4	3				5	6	1
t116	2		5			5	5		3	2				1	4	2		1	4	5	5	3	7	5			4	4	5	5					5	2

Fig. 4. The priority corresponding to the tasks in each LEO satellite at each time slot

The number of tasks completed per LEO under each time slot is shown in Fig. 6. It can be seen from the figure that the number of tasks completed is lower than that of the higher priority tasks when the tasks are of the same size, as the rate of lower priority task requests is lower than the rate of higher priority tasks.

The cost of LEO under each time slot is shown in Fig. 7. As can be seen from the figure, the cost is proportional to the priority, with higher priority tasks

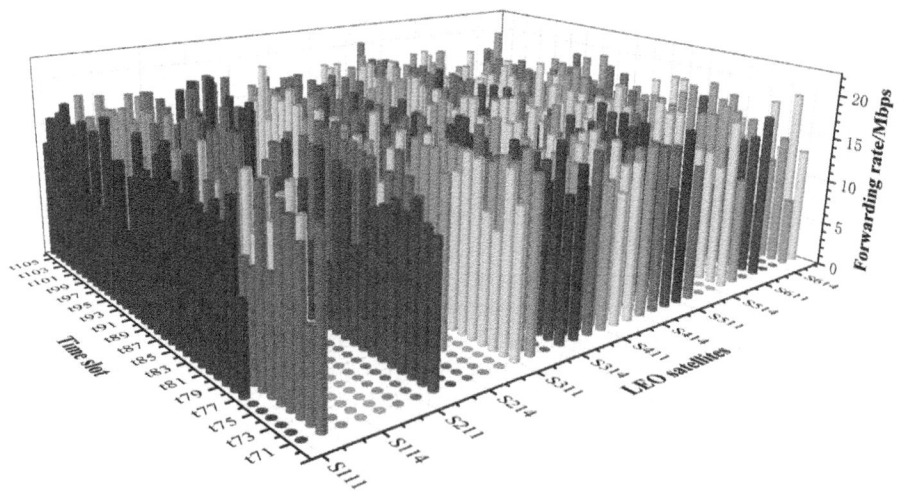

Fig. 5. Comparison of transmission rates for different priority applications at each time slot

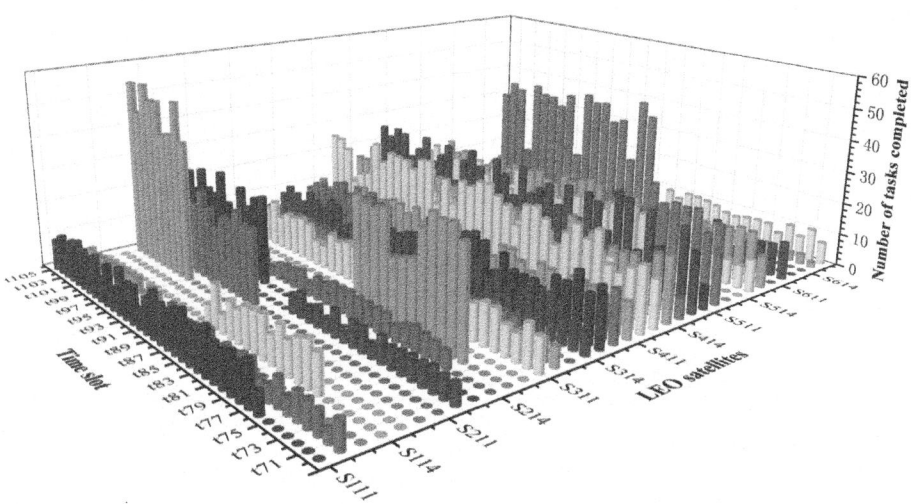

Fig. 6. Comparison of transmission rates for different priority applications at each time slot

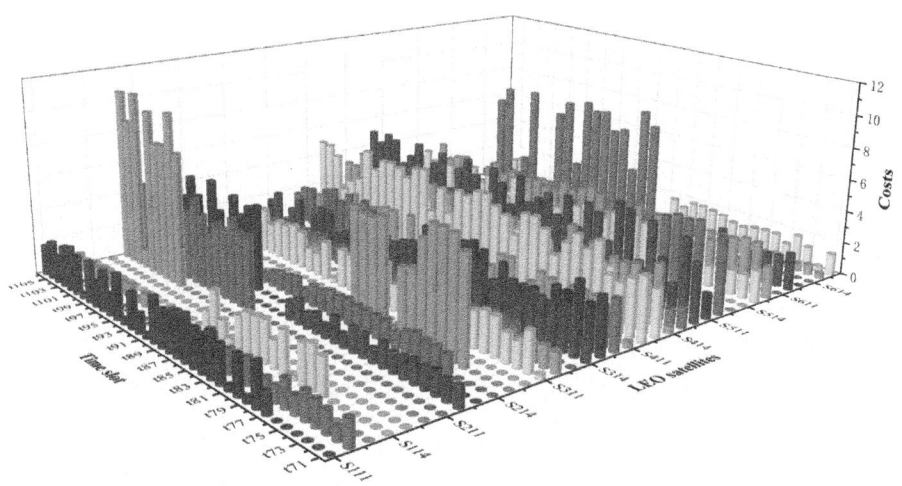

Fig. 7. Cost per LEO under each time slot

requesting a higher forwarding rate, corresponding to higher cost, and obtaining a higher task completion efficiency.

The total transmission rate allocated out by G1 under each time slot is shown in Fig. 8. It can be seen from the figure that the maximum resource utilisation of G1 is about 90% and the minimum is 60%, this is because the total amount allocated under different time slots is related to the priority of the tasks applied for forwarding under the current time slot, the greater the sum of priority the higher the utilisation and vice versa the lower the utilisation.

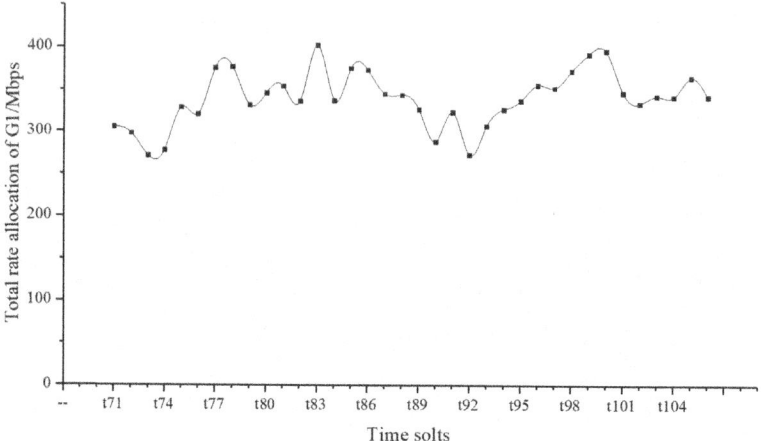

Fig. 8. The total transmission rate allocated out by G1 under each time slot

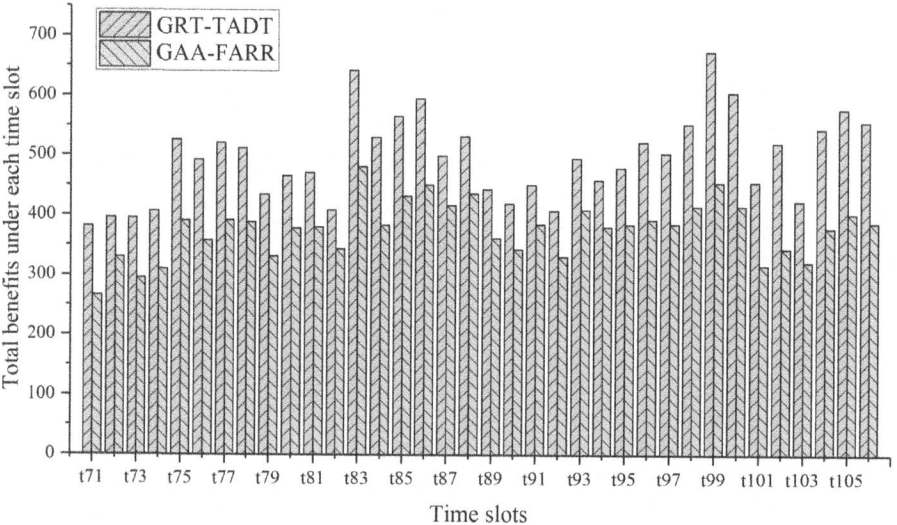

Fig. 9. Comparison of benefits under each time slot

Finally, the benefit comparison between the GRT-TADT scheme and the GAA-FARR strategy under each time slot is analyzed, as shown in Fig. 9. From the figure, it can be seen that the gains of the GRT-TADT scheme proposed in this paper are larger than those of the GAA-FARR strategy, with the largest difference at time slot t98, the GRT-TADT scheme being 48.2% higher than the GAA-FARR strategy, which is because the GAA-FARR strategy does not consider the impact of task level on the benefits, and all tasks are transmitted at the same rate.

6 Conclusion

To address the problem of real-time remote sensing data transmission service, this paper proposes a real-time remote sensing GEO relay transmission scheme based on task awareness, setting different forwarding rates for tasks with different priorities, aiming to maximize GEO benefits. Due to the dynamic change of LEO topology, the LEOs applying to GEO for forwarding tasks within each time slot are also different, therefore, this paper analyses the LEO identity, the resource allocation of GEO under different task priorities and the benefit, and compared with the GAA-FARR strategy in terms of benefit, and the experimental results prove the effectiveness of the scheme proposed in this paper.

References

1. Araniti, G., et al.: Contact graph routing in DTN space networks: overview, enhancements and performance. IEEE Commun. Mag. **53**(3), 38–46 (2015)
2. Bertsekas, D.P.: Nonlinear programming. J. Oper. Res. Society **48**(3), 334–334 (1997)
3. Chen, J., et al.: An efficient scheme for in-orbit remote sensing image data retrieval. Futur. Gener. Comput. Syst. **150**, 103–114 (2024)
4. Chen, J., et al.: A remote sensing data transmission strategy based on the combination of satellite-ground link and geo relay under dynamic topology. Futur. Gener. Comput. Syst. **145**, 337–353 (2023)
5. Deng, C., Guo, W., Hu, W., Zhu, W., Zhou, B.: Algorithm for the lightpath reservation provisioning of data relay services in a geo network. J. Optical Commun. Network. **9**(8), 658–668 (2017)
6. Deng, R., Di, B., Song, L.: Pricing mechanism design for data offloading in ultra-dense leo-based satellite-terrestrial networks. In: 2019 IEEE Global Communications Conference (GLOBECOM), pp. 1–6. IEEE (2019)
7. Du, J., Jiang, C., Wang, J., Ren, Y., Yu, S., Han, Z.: Resource allocation in space multiaccess systems. IEEE Trans. Aerosp. Electron. Syst. **53**(2), 598–618 (2017)
8. Jamilkowski, M.L., Grant, K.D., Miller, S.W.: Support to multiple missions in the joint polar satellite system (jpss) common ground system (cgs). In: AIAA SPACE 2015 Conference and Exposition, p. 4469 (2015)
9. Kelly, F.P., Maulloo, A.K., Tan, D.K.H.: Rate control for communication networks: shadow prices, proportional fairness and stability. J. Oper. Res. Society **49**(3), 237–252 (1998)
10. Lillesand, T., Kiefer, R.W., Chipman, J.: Remote Sensing and Image Interpretation. John Wiley & Sons, New York (2015)
11. Palomar, D.P., Chiang, M.: A tutorial on decomposition methods for network utility maximization. IEEE J. Sel. Areas Commun. **24**(8), 1439–1451 (2006)
12. Qi, K., Hu, Y., Li, S., Zhai, W., Cheng, C.: An improved method for the unique code of spatial entity based on global subdivision grid. In: 2017 IEEE International Geoscience and Remote Sensing Symposium (IGARSS), pp. 6067–6070. IEEE (2017)
13. Roddy, D.: Satellite communications. McGraw-Hill Education (2006)
14. Tong, X., et al.: Normalized projection models for geostationary remote sensing satellite: a comprehensive comparative analysis (january 2019). IEEE Trans. Geosci. Remote Sens. **57**(12), 9643–9658 (2019)

15. Wang, E., Li, H., Zhang, S.: Load balancing based on cache resource allocation in satellite networks. IEEE Access **7**, 56864–56879 (2019)
16. Wang, L., Jiang, C., Kuang, L., Wu, S., Huang, H., Qian, Y.: High-efficient resource allocation in data relay satellite systems with users behavior coordination. IEEE Trans. Veh. Technol. **67**(12), 12072–12085 (2018)
17. Xu, Q., Su, Z., Lu, R., Yu, S.: Ubiquitous transmission service: Hierarchical wireless data rate provisioning in space-air-ocean integrated networks. IEEE Trans. Wireless Commun. **21**(9), 7821–7836 (2022)
18. Xu, X., Zhao, H., Liu, C., Wang, Q., Wang, S.: Resource management of geo relays for real-time remote sensing. Peer-to-Peer Networking and Applications **14**, 3333–3348 (2021)
19. Zhang, H., Song, Y., Han, C., Zhang, L.: Remote sensing image spatiotemporal fusion using a generative adversarial network. IEEE Trans. Geosci. Remote Sens. **59**(5), 4273–4286 (2020)

Review of Image Encryption Based on Bibliometrics and Visualized Analysis

Zhaoyang Liu[1,2,3] and Ru Xue[1,2,3](✉)

[1] School of Information Engineering, Xizang Minzu University, Xianyang 712082, Shaanxi, China
rxue@xzmu.edu.cn
[2] Key Laboratory of Optical Information Processing and Visualization Technology of Tibet Autonomous Region, Xianyang 712082, Shaanxi, China
[3] Xizang Cyberspace Governance Research Center, Xianyang, China

Abstract. Image encryption is a key technology for protecting image confidentiality and integrity in various applications. However, there is a lack of comprehensive and quantitative overview of the relevant publications in this field. This paper aims to fill this gap and explore the evolution and research frontier of image encryption. We collected papers from the WOS core database from 2005 to 2022 and used R programming language and VOSviewer to conduct bibliometric analysis and visualization. We analyzed the annual output, publication countries and institutions, journal sources, authors, and influential papers. We also visualized the author relationship, institutional cooperation relationship, keyword co-occurrence network, etc. We identified the research hotspots as: 1) chaotic systems; 2) optical encryption; 3) reversible information hiding; 4) deep learning; 5) visual encryption; 6) quantum image encryption. We found that: 1) the literature volume shows an overall upward trend; 2) China has a leading position and close cooperation with many countries; 3) quantum image encryption and deep learning image encryption are emerging topics with large development space. This paper provides valuable references for researchers and practitioners in the area of image encryption.

Keywords: Image encryption · Bibliometrics analysis · visualization analysis · VOSviewer

1 Introduction

In modern society, with the rapid development of computer technology and the Internet, various forms of information are facing data leakage and security issues, making it particularly important to protect information security. In the field of information protection, image encryption is a very important technical means, which can convert sensitive information into an unreadable format and can only be read and used by authorized personnel, thus protecting the confidentiality and integrity of images. Today, image encryption technology is widely used by individuals, businesses, government agencies, and military fields. In the personal domain, image encryption technology can protect the security of personal photos, identity cards, and driver's licenses. In commercial applications, image encryption can prevent the leakage of corporate secrets and ensure the security of business information. In the government and military fields, image encryption can prevent national secrets from being obtained by the enemy and maintain national security. Therefore, image encryption technology has been widely used in privacy and confidentiality protection.

Many scholars have conducted a lot of research in the field of image encryption. However, most review articles start from a single perspective: some scholars focus on color image encryption [1], some scholars focus on medical applications [2, 3], some scholars focus on chaotic encryption [4, 5], and some scholars focus on deep learning encryption [6]. Although these existing articles are valuable, they have failed to objectively reveal the overall picture of this field, nor have they appropriately described the key issues of the current research hotspots, cooperation networks, and development trends in the field of image encryption. Moreover, in traditional review articles, it is difficult to effectively organize, summarize and quantitatively analyze the development of a specific field over a long period of time. Image encryption is an interdisciplinary research field that covers mathematics, optics, physics, computer science, and other disciplines.

However, the problems mentioned above can be easily solved by using bibliometric analysis [7], which is a quantitative analysis of books, articles or other publications. In recent years, bibliometric analysis has been widely used in many fields, such as natural science [8, 9], information science [10, 11], and humanities and social sciences [12, 13]. This can help interested scholars to understand the current situation, evolution, and development trends of the field. Therefore, the structure of this paper is as follows: first, we introduce the data source and methods. Secondly, we conduct bibliometric analysis from the aspects of annual paper output, publishing countries and institutions, journal sources, authors, and the most influential papers. Then, we use VOSviewer to summarize the academic relations, research hotspots, and emerging trends in the field through cluster analysis of country and institution cooperation, author co-authorship network, and keyword network. Finally, we summarize the development trajectory and future direction of image encryption, providing meaningful guidance for scholars and research institutions interested in the field of image encryption.

Overall, the advantages of our study are summarized as follows:

1. We have conducted bibliometric analysis on the literature in the field of image encryption. Bibliometric analysis is a quantitative analysis method for literature review, which can grasp the hotspots and trends of the research field from a comprehensive perspective, and it is more comprehensive and objective than relying solely on literature review.
2. We have used the R programming language and the visualization analysis tool VOSviewer to conduct bibliometric analysis and co-occurrence network analysis on relevant literature, revealing the development of research directions in image encryption in an intuitive way, providing valuable and pioneering references for related researchers and practitioners.
3. We have summarized the theoretical development trajectory of image encryption over the past 18 years and the current research hotspots and future development trends in

the field of image encryption, providing a meaningful overview for scholars interested in studying image encryption.

2 Data Source and Methodology

2.1 Data Sources and Pre-processing

Web of Science (WOS) is an important database for accessing academic information worldwide, including information in natural sciences, social sciences, arts, and humanities, and contains nearly 9,000 of the most prestigious high-impact research journals and more than 12,000 academic conferences in various fields [14]. Among them, the Web of Science core database includes more than 8,700 authoritative and high-impact academic journals globally, covering fields such as natural sciences, engineering and technology, biomedical, social sciences, arts, and humanities [15]. It is a multidisciplinary, large-scale comprehensive citation index database built based on the web and a dynamically updated digital research environment integrated with core academic information resources. In this paper, the Web of Science core database is used as the data source, with "image encryption" selected as the search keyword. The time span was limited to 2005–2022, and the types of literature included research articles, review papers, and conference papers, while editorials, corrections, and book chapters were excluded. A total of 7,420 related articles were selected as research samples.

2.2 Research Methods

This article employs bibliometric analysis and knowledge graph analysis as the main research methods. Bibliometrics is a research method that reflects the development trends of a discipline by statistically analyzing aspects such as the number of published literature, authors, countries, and journals [16]. Knowledge graph displays a visual representation of the complex relationships between the development process of a scientific knowledge and the intersection, interaction, and evolution of knowledge units, which intuitively reflect co-operation relationships, research hotspots, and research frontiers [17]. Therefore, the combination of both can not only highlight the trends of scientific development, but also reveal new topics with development prospects for promoting the development of science and technology. Image encryption is one of the important branches in the field of image processing. With the continuous development of computer technology, encryption techniques are constantly being upgraded, and encryption algorithms have become increasingly complex to improve image security. Therefore, bibliometric and knowledge graph analysis of literature in the field of image encryption is of significant importance for future research and development.

In the software used for measuring and analyzing scientific literature data, VOSviewer [18], CiteSpace [19], and UCINET [20] have been widely used in the global information science field due to their scientific and effective advantages in practical applications [21]. Among them, VOSviewer software based on the principle of literature co-citation can visualize the structure, patterns, and distribution of scientific knowledge, and can reveal the research situation of a certain discipline in a certain period, and can also be used to predict research hotspots and development trends in related fields, and can comprehensively reflect the current research status and dynamic changes. Especially when there are many keyword co-citations, its analytical ability is strong. CiteSpace has the advantage of rich visualization and the ability to extract network relationships and research focuses. UCINET is one of the most popular social network analysis software and is often used for one-dimensional and two-dimensional data analysis. In addition, the R programming language also provides software packages specifically developed for bibliometric and scientometric research. R language has statistical computing and rapid visualization advantages in bibliometric analysis [22], making it an effective and flexible tool. Compared with complex bibliometric analysis software, it can make up for the deficiency of researchers unable to build a complete workflow for literature analysis and improve work efficiency, and can handle large-capacity and high-repetition computing tasks.

Therefore, this article uses R programming language bibliometric analysis and VOSviewer analysis tool to conduct statistical analysis and draw a clear relationship knowledge graph, including quantitative analysis of year publish volume, country, author, institution, and journal, and knowledge graph analysis of the current situation, research hotspots, and development background of the image encryption field, providing a reference for future research in this field. The bibliometric analysis and visualization process of this study are shown in Fig. 1.

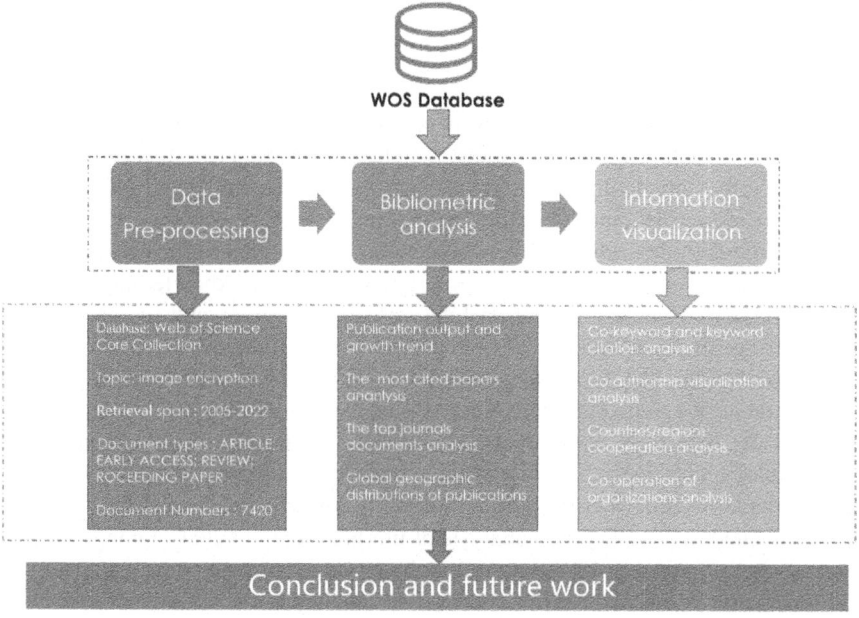

Fig. 1. Bibliometric analysis and visualization process

All data in this study were obtained from the WoS core database and went through data preprocessing. Firstly, bibliometric analysis using the R programming language was conducted to perform statistical analysis on the annual publication volume, publishing countries and institutions, highly cited literature, and other aspects of all search results, thereby reproducing the development process of the image encryption research field. Next, VOSviewer was used to cluster analysis of authors, countries, institutions, keywords, and other aspects using co-occurrence network analysis methods in order to gain a more intuitive understanding of the current research hotspots and future development trends.

3 Results

3.1 Annual Publication Outputs Analysis

According to the bibliometric theory, the statistics of temporal distribution of the number of literature resources can reflect the research level and development of the field to a certain extent. In this paper, we present the annual distribution of articles related to image encryption by statistically analyzing 7420 documents in the WOS core database, as shown in Fig. 2.

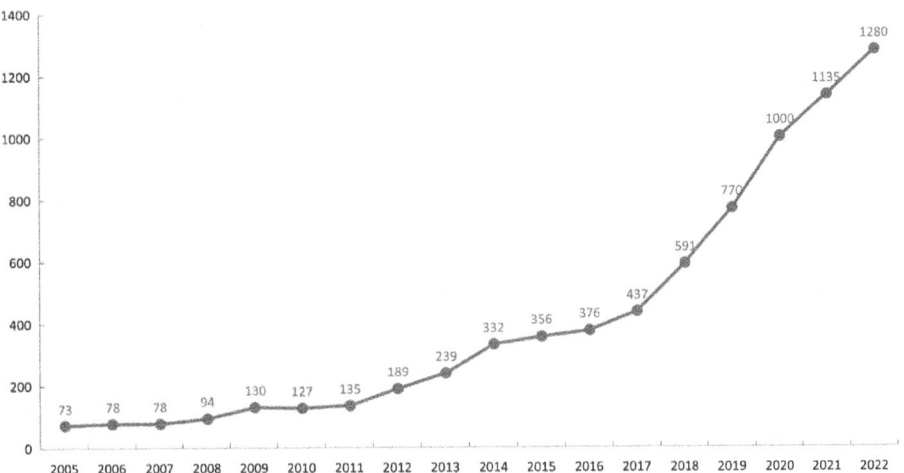

Fig. 2. Annual distribution of published literature in 2005–2022

As shown in the figure, the number of publications in the field of image encryption has generally shown an upward trend year by year. A total of 7,420 related literature have been published from 2005 to 2022, which can be divided into two phases: a stable growth period from 2005 to 2017, and a high-productivity period from 2018 to 2022. This indicates that image encryption technology has received more and more attention over time. In recent years, especially after 2018, the number of image encryption literature has shown an explosive growth, indicating that people's attention to image encryption technology is increasing. This may be related to the development of China's network security strategy in 2018, which put forward policies, work tasks, and measures for network security in key areas.

3.2 The Publication of the Countries and Institutions Analysis

The number of national publications reflects to some extent the importance of the research direction for the country, and the number of publications and cited scientific papers is an important indicator of a country's scientific power. In this paper, by analyzing the national sources of published papers, the top ten countries in the field of image cryptography have been derived from the literature published, as shown in Table 1.

Table 1. Top 10 countries with the highest number of publications

Country	Articles	SCP	MCP	TC	Average Article Citations
CHINA	3826	3241	585	103603	27.10
INDIA	911	911	99	15363	16.90
EGYPT	201	94	107	3722	18.50
KOREA	200	120	80	3105	15.50
PAKISTAN	184	97	87	3159	17.20
IRAN	152	130	22	4954	32.60
USA	130	77	53	5936	45.70
TURKEY	98	78	20	2234	22.80
SINGAPORE	75	59	16	2772	37.00
FRANCE	75	41	34	2473	33.00

Note: SCP, single-country publications; MCP, multiple-country publications

In terms of the research countries in the field of image encryption, China is the most active country with 3826 publications, accounting for 54.5% of the total literature. India is the next most active country, with 911 publications, accounting for 15.3% of the total literature. Other active countries/regions include egypt, korea, pakistan, etc. These countries also have great enthusiasm for research in the field of image encryption and have achieved certain results.

In addition, the higher average article citations are in the US, Singapore and France, respectively, indicating the strong academic influence of these three countries. However, the average article citation of Chinese papers is relatively low at 27.10, which indicates that the quality of Chinese papers needs further improvement. Specifically, Table 2 lists the 10 institutions with the highest publication output.

As can be seen from the data, menoufia university has the highest number of papers published among the top 10 institutions, all of which have made significant contributions to this research. It is worth noting that 9 of these institutions are in China, reflecting the outstanding research results achieved in China in the field of image encryption.

3.3 High-Quality Journals Analysis

A study of the distribution of journal sources in a research field is an effective way to understand the high-quality, core journals in that field. In this paper, the top five journals

Table 2. Top 10 institutions with the highest number of publications

Rank	Institution	Publications	Country
1	MENOUFIA UNIV	186	Egypt
2	HARBIN INST TECHNOL	172	China
3	DALIAN UNIV TECHNOL	167	China
4	SHENZHEN UNIV	151	China
5	DALIAN MARITIME UNIV	149	China
6	CHONGQING UNIV	140	China
7	NORTHEASTERN UNIV	130	China
8	NANCHANG UNIV	125	China
9	SHANDONG UNIV	114	China
10	HENAN UNIV	113	China

were ranked by the number of journals in which the papers were published, as shown in Fig. 3.

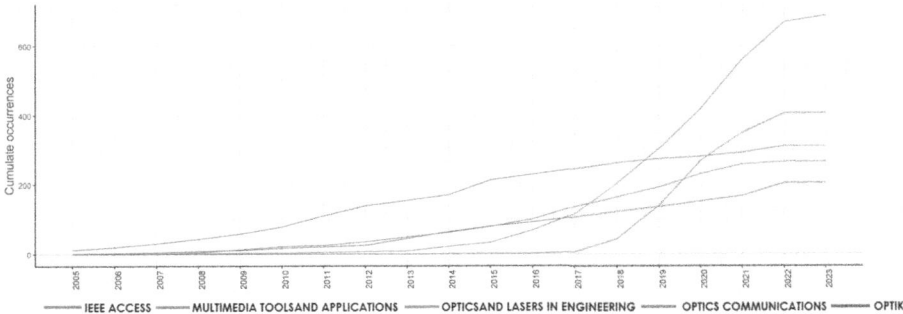

Fig. 3. Trends in annual publications of different journals

As can be seen from the figure, all journal publications are increasing year by year, with multimedia tools and applications journals having the highest number of publications, much higher than other journals. In addition, a larger proportion of papers are published in optical journals. It is noteworthy that the number of publications in IEEE access journals came later after 2020, from the fifth to the second rank, which shows that multimedia tools and applications and IEEE access are the more popular journals in recent years.

3.4 Highly Cited Scholar Representatives Analysis

Highly cited scholars represent the status in this field and reflect the high point of national knowledge innovation. To address this point, the top ten highly cited authors in the field of image cryptography were counted in this paper, as shown in Table 3.

Table 3. The 10 most influential authors in the field

Author	H-index	G-index	TC	NP	Country
WANG XY	52	97	10737	245	China
LIU ZJ	35	54	3029	78	China
ZHANG YS	34	56	3234	82	China
ZHOU NR	33	55	3236	55	China
LIU ST	32	52	2801	56	China
CHEN W	29	52	2830	67	China
ZHAO DM	27	43	2069	43	China
ZHANG Y	26	45	2061	62	China
ABD EL-LATIF AA	25	45	2106	48	Egypt
CHEN XD	25	48	2354	50	China

Note: h-Index represents the importance and the influence of the accumulated research; the g-Index represents the derivative index of the h-index; TC stands for total citations; NP represents the number of publications

The h-index is commonly used to assess the number and level of scholarly output of researchers, and the g-index reflects the average number of citations of articles. From the data, WANG XY has the highest number of publications, with an h-index of 52, a g-index of 97, and a total number of citations of 10737, which is much higher than the other authors, showing his high innovative ability and high influence.

3.5 High-Influence Literature Analysis

A paper is an important aspect of academic research output, and the more times a paper is cited, the higher the quality of the paper and the greater its impact in the field. By analyzing the literature citations in the field of image encryption, we can understand the research hotspots and frontier directions in the field. In this paper, the top ten papers with high impact in the field of image encryption are listed in Table 1 by ranking the number of citations of the articles.

From the data, it can be seen that these papers mainly involve research on image encryption algorithms based on chaotic, optical, DNA coding and reversible information hiding, which are of high research value. Among them, the articles on chaotic image encryption have the highest citation frequency of 762, and the articles on one-dimensional [24], and two-dimensional chaotic encryption [25] have higher average annual citations than other articles, which can be seen that chaotic image encryption still receives a high level of attention. It is worth noting that most of the top ten cited articles focus on chaotic image encryption, and it can be seen that chaos has a profound impact on the field of image encryption (Table 4).

Table 4. The top 10 High-influence literature

Rank	Publication	Author	Citations			TC	CY
			2020	2021	2022		
1	Image encryption using chaotic logistic map	NK et al.	71	56	45	762	42.33
2	Known-plaintext attack on optical encryption based on double random phase keys	Zhang et al.	52	36	31	579	32.17
3	A new 1D chaotic system for image encryption	Zhou et al.	101	94	83	541	54.1
4	Lest We Remember: Cold-Boot Attacks on Encryption Keys	JA et al.	38	51	42	539	35.93
5	A novel colour image encryption algorithm based on chaos	Wang et al.	80	73	65	536	44.67
6	Reversible Data Hiding in Encrypted Image	Zhang et al.	69	75	78	535	41.15
7	2D Sine Logistic modulation map for image encryption	Hua et al.	77	101	83	473	52.56
8	A symmetric image encryption algorithm based on mixed linear-nonlinear coupled map lattice	Zhang et al.	84	88	47	469	46.9
9	Image encryption using DNA complementary rule and chaotic maps	Liu et al.	77	67	68	463	38.58
10	A chaos-based symmetric image encryption scheme using a bit-level permutation	Zhu et al.	66	54	37	450	34.62

Note: TC: Total Citations; CY: Citations per year

3.6 Visual Analysis of Author Relationships

Authors are the main body of scientific research. The analysis of structural characteristics of authors and their collaboration networks can reflect the core group of authors and their collaborative relationships in the field. In order to analyze team collaboration in the field of image encryption, we construct a map of author collaboration networks by

VOSviewer. When creating the author data based on the collaborative author map, the threshold was set to 4 in order to present the authors who have published on the topic of image encryption in the network, and the results are shown in Fig. 4.

Fig. 4. Co-occurrence density view of published authors

The more than 10 different color categories seen in the figure indicate the collaborative clusters of authors, where the size of the circle represents the status of the author and the lines between the authors indicate their collaborative links. It can be seen that the best researchers in the field of image encryption are wang xingyaun, zhan yushu, chen wen, etc. They all have their own research teams and have collaborative relationships with other researchers.

3.7 Visual Analysis of Countries Cooperation

Based on the bibliographic data collected from the WoS core, a visual mapping of the country collaboration network was created using VOSviewr, and the minimum issuance threshold for a country was set to 10 during the mapping process, and the results are shown in Fig. 5.

In the diagram, the size of the circles indicates the number of organizations and the thickness of the lines indicates the closeness of the relationships. The six different colors in the figure can distinguish six scientific camps of image encryption research, indicating that there are different degrees of cooperation among countries. For example, there is close cooperation between China and the United States, Australia, and Canada, and the organization of Chinese scholars is the largest.

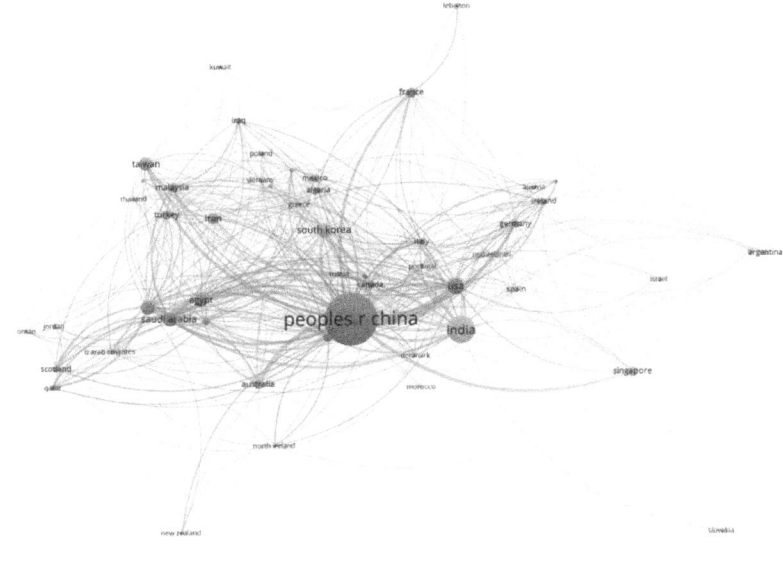

Fig. 5. Country cooperation co-occurrence network analysis

3.8 Visual Analysis of Organizations Cooperation

In order to examine the representative organizations and their cooperative relationships in the research field, we constructed a network mapping of institutional cooperative relationships, and when the threshold value was 5, a total of 576 organizations were screened, and the results are shown in Fig. 6.

As can be seen in the figure, the 184 representative organizations are divided into at least eight clusters, indicated by eight colors. The size of the circle indicates the number of publications, and the line between two nodes indicates the academic connection between the two organizations. The shorter the line, the stronger the relationship. Therefore, the research institutions in the direction of image encryption are mainly concentrated in Chinese universities, such as Nanchang University, University of Science and Technology Beijing, and Harbin University of Science and Technology, showing the enthusiasm of Chinese researchers in the field of image encryption. In addition, China and Singapore, Egypt and India are also very close to each other.

3.9 Visual Analysis of High Frequency Keywords

Keywords are the condensation of an article's content, and high-frequency keywords are words with high co-occurrence frequency. In this paper, keywords in the field of image encryption are counted by VOSviewer, and the frequency is set greater than or equal to 10 times as high-frequency keywords, and the keyword co-occurrence network mapping is drawn to show the closeness of connection between research topics, which can also reflect the hot topics in the research field. The results are shown in Fig. 7.

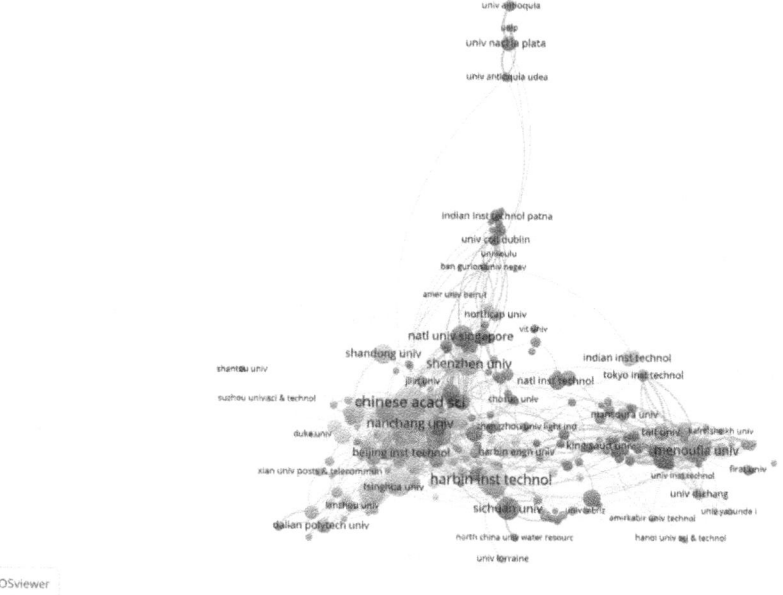

Fig. 6. Organization cooperation co-occurrence network analysis

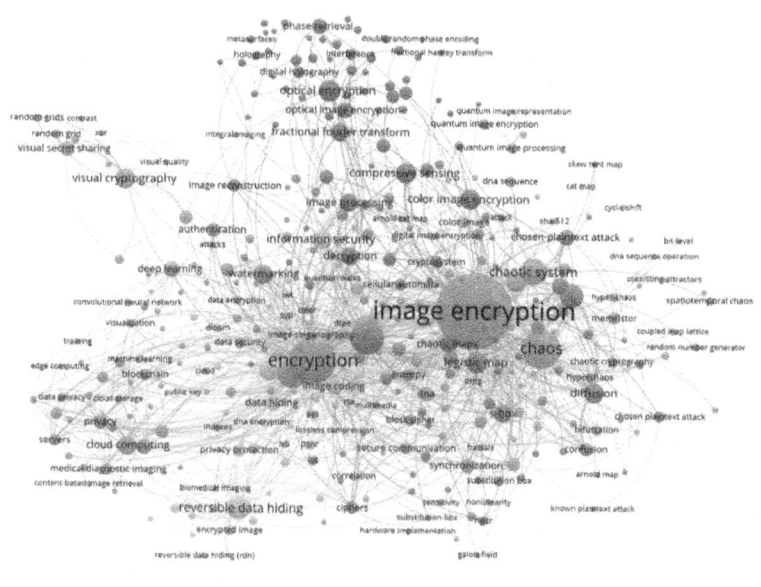

Fig. 7. Co-occurrence Analysis of Keywords of Image Encryption

In the figure, each node represents a keyword, and the higher the frequency of its occurrence, the larger the node. 1 color represents a cluster, and the cluster is composed of keywords with high co-occurrence frequency, and the connecting line in the figure represents that the 2 keywords connected have at least 1 co-occurrence in 1 literature, and the higher the frequency of keyword co-occurrence, the thicker the line is. Through the keywords in each cluster, six research hotspots of image encryption can be summarized, which are: 1 chaotic system; 2 optical encryption; 3 reversible information hiding; 4 deep learning; 5 visual encryption; 6 quantum image encryption. Among them, such as chaotic systems, optical image encryption, reversible information hiding and visualization encryption have a higher frequency of co-occurrence and are more research directions in the field.

In addition, the keywords with higher frequency in a certain period of time may be the research hotspots or research focus in a certain field at that time. The high frequency keywords obtained from the co-occurrence analysis of keywords by the visualization analysis software can reveal the thematic structure and hot issues of a research field, which is more conducive to the comprehensive analysis of a research field, and the visualization results are shown in Fig. 8.

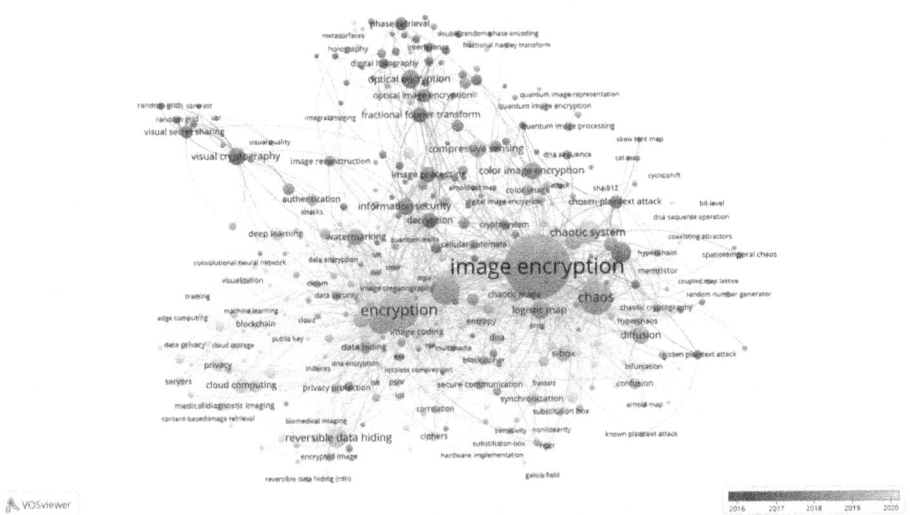

Fig. 8. Co-occurrence Analysis of Keywords of Image Encryption

Where the node size indicates the frequency of keyword occurrences. The thickness of the connecting line indicates the intensity of co-occurrence between keywords, and the color indicates the time node of occurrence. As can be seen from the figure, the proliferation of phrases such as "deep learning-based encryption", "reversible encryption" and "quantum encryption" in recent years implies that the research on digital image encryption is shifting from traditional encryption methods to more advanced and sophisticated encryption techniques. This shift may be a response to the increasing sophistication of digital image attacks. The development trend of image encryption is from optical image encryption to chaos to combined information hiding techniques to quantum image encryption to deep learning image encryption, where quantum image encryption [26] and deep learning image encryption [27] are in the early stage of development and will become a hot spot for future research. In addition, for cloud computing, image encryption applications in medical diagnosis are also emerging gradually [28].

4 Conclusions and Outlook

This paper provides a broader perspective on the scientific study of image encryption and reveals the trends in this field. We analyzed 7420 papers related to image encryption from the WOS core collection from 2005–2022 by bibliometric methods. The findings show that the number of papers published in the direction of image encryption shows a gradual increase, experiencing two stages of steady growth and accelerated growth. Especially after 2017, the number of literature publications increased significantly. Nine of the top 10 productive institutions are from China, reflecting the importance China attaches to this field. In addition, the analysis of national and institutional collaborative co-occurrence networks shows that image encryption has been studied by scholars from all over the world, indicating that the research on this topic is a global effort. Among them, China is a prominent leader among the countries in the world and has established close collaborative relationships with many countries or regions. The United States, India, Iran, and South Korea have also accomplished fruitful work and made great contributions. Second, several universities and research institutions play a crucial role in the collaborative network, and the most important core authors are Eastern scientists. Image encryption research is mainly concentrated in universities, such as menoufia university, Nanchang University, University of Science and Technology Beijing, Harbin University of Science and Technology, etc. From the perspective of cooperative network, the image encryption research shows the characteristics of small concentration and large dispersion. The analysis of authors reveals that many authors tend to collaborate with a small group of collaborators, producing several major author groups such as wang xingyaun, zhan yushu, chen wen. According to the analysis of high quality journals, Multimedia Tools and Applications is an influential and productive journal with the fastest rising number of articles and citations. In recent years, IEEE Access journals have attracted the attention of researchers and advanced the process of publishing image encryption articles, attracting many scholars around the world.

What's more, the research hotspots in the field of image encryption have gradually shifted from high concentration to diversity, and the visual analysis of keyword co-occurrence clustering by VOSviewer identifies research trends and hotspots and reveals the future development direction. From the high-frequency keywords, the research hotspots of image encryption are mainly distributed in: 1 chaotic system; 2 optical encryption; 3 reversible information hiding; 4 deep learning; 5 visualization encryption; 6 quantum image encryption. Among them, quantum image encryption and deep learning image encryption are in the early stage of development and have a large development space, so we can pay more attention to these directions, which will be promising directions in the field of image encryption.

Overall, this paper provides a clearer and deeper understanding of the research area of image encryption using R programming language bibliometrics and VOSviewer analysis tool, and reveals the development trend of the image encryption field. We hope that these results will benefit researchers working in this field to help researchers identify knowledge gaps in the field, clarify their own future research directions, and develop new and innovative algorithms to provide better security for digital images.

Funding. This project is supported in part by the National Natural Science Foundation of China: 62262062, the major programs incubation plan of Xizang Minzu University: 22MDZ03, and the Research Team Project for Xizang-related Network Information Content and Data Security (No. 324042000709).

Competing Interests. The authors declare that they have no known competing financial interests or personal relationships that could have appeared to influence the work reported in this paper.

Authors Contribution Statement. Author Liu conceived and designed the study, collected and analyzed the data, and wrote the manuscript. Author Xue contributed to the data collection and analysis, provided critical revisions to the manuscript, and approved the final version. All authors have read and approved the final manuscript.

Ethical and Informed Consent for Data Used. The data used in this study were obtained from a publicly available dataset, which was collected in accordance with ethical principles and informed consent procedures. The original study received ethical approval from the Institutional Review Board (IRB) prior to data collection. The participants provided informed consent for the collection and use of their data for research purposes. All identifying information was removed from the public dataset to protect the privacy and confidentiality of the participants. The current analysis was conducted in compliance with the ethical standards set forth by the IRB.

Data Availability and Access. The data that support the findings of this study are available from the corresponding author upon reasonable request.

References

1. Ghadirli, H.M., Nodehi, A., Enayatifar, R.: An overview of encryption algorithms in color images. Signal Process. **164**, 163–185 (2019)
2. Singh, A.K.: A survey of image encryption for healthcare applications. Evol. Intell. 1–18 (2022)
3. Pavithra, V., Jeyamala, C.: A survey on the techniques of medical image encryption. In: 2018 IEEE International Conference on Computational Intelligence and Computing Research (ICCIC), pp. 1–8. IEEE, December 2018
4. Özkaynak, F.: Brief review on application of nonlinear dynamics in image encryption. Nonlinear Dyn. **92**(2), 305–313 (2018)
5. Kumar, M., Saxena, A., Vuppala, S.S.: A survey on chaos based image encryption techniques. In: Multimedia Security Using Chaotic Maps: Principles and Methodologies, pp. 1–26 (2020)
6. Bao, Z., Xue, R.: Survey on deep learning applications in digital image security. Opt. Eng. **60**(12), 120901 (2021)
7. He, X., Wu, Y., Yu, D., Merigó, J.M.: Exploring the ordered weighted averaging operator knowledge domain: a bibliometric analysis. Int. J. Intell. Syst. **32**(11), 1151–1166 (2017)
8. Gao, Q., et al.: The top 100 highly cited articles on osteoporosis from 1990 to 2019: a bibliometric and visualized analysis. Arch. Osteoporos. **15**, 1–11 (2020)

9. Xie, L., Chen, Z., Wang, H., Zheng, C., Jiang, J.: Bibliometric and visualized analysis of scientific publications on atlantoaxial spine surgery based on Web of Science and VOSviewer. World Neurosurg. **137**, 435–442 (2020)
10. Yu, D., Xu, Z., Wang, X.: Bibliometric analysis of support vector machines research trend: a case study in China. Int. J. Mach. Learn. Cybern. **11**, 715–728 (2020)
11. Dhamija, P., Bag, S.: Role of artificial intelligence in operations environment: a review and bibliometric analysis. TQM J. **32**, 869–896 (2020)
12. Shi, Y., Blainey, S., Sun, C., Jing, P.: A literature review on accessibility using bibliometric analysis techniques. J. Transp. Geogr. **87**, 102810 (2020)
13. Garrigos-Simon, F.J., Narangajavana-Kaosiri, Y., Narangajavana, Y.: Quality in tourism literature: a bibliometric review. Sustainability **11**(14), 3859 (2019)
14. Mongeon, P., Paul-Hus, A.: The journal coverage of Web of Science and Scopus: a comparative analysis. Scientometrics **106**, 213–228 (2016)
15. Liu, W.: The data source of this study is Web of Science Core Collection? Not enough. Scientometrics **121**(3), 1815–1824 (2019)
16. de Oliveira, O.J., da Silva, F.F., Juliani, F., Barbosa, L.C.F.M., Nunhes, T.V.: Bibliometric method for mapping the state-of-the-art and identifying research gaps and trends in literature: an essential instrument to support the development of scientific projects. In: Scientometrics Recent Advances. IntechOpen (2019)
17. Cao, Y., Qi, F., Cui, H., Yuan, M.: Knowledge domain and emerging trends of carbon footprint in the field of climate change and energy use: a bibliometric analysis. Environ. Sci. Pollut. Res. **30**, 1–18 (2022)
18. Tupan, T.: Pemetaan Bibliometrik Dengan Vosviewer Terhadap Perkembangan Hasil Penelitian Bidang Pertanian Di Indonesia. Visi Pustaka: Buletin Jaringan Informasi Antar Perpustakaan **18**(3), 217–230 (2016)
19. Chen, C.: CiteSpace II: detecting and visualizing emerging trends and transient patterns in scientific literature. J. Am. Soc. Inform. Sci. Technol. **57**(3), 359–377 (2006)
20. Johnson, J.D.: UCINET: a software tool for network analysis (1987)
21. Oguntimilehin, A., Ademola, E.O.: A review of big data management, benefits and challenges. Rev. Big Data Manage. Benefits Challenges **5**(6), 1–7 (2014)
22. Guler, A.T., Waaijer, C.J., Mohammed, Y., Palmblad, M.: Automating bibliometric analyses using Taverna scientific workflows: a tutorial on integrating Web Services. J. Informet. **10**(3), 830–841 (2016)
23. Guleria, D., Kaur, G.: Bibliometric analysis of ecopreneurship using VOSviewer and RStudio Bibliometrix, 1989–2019. Library Hi Tech **39**(4), 1001–1024 (2021)
24. Pareek, N.K., Patidar, V., Sud, K.K.: Image encryption using chaotic logistic map. Image Vis. Comput. **24**(9), 926–934 (2006)
25. Hua, Z., Zhou, Y., Pun, C.M., Chen, C.P.: 2D Sine Logistic modulation map for image encryption. Inf. Sci. **297**, 80–94 (2015)
26. Zhang, J., Huang, Z., Li, X., Wu, M., Wang, X., Dong, Y.: Quantum image encryption based on quantum image decomposition. Int. J. Theor. Phys. **60**, 2930–2942 (2021)
27. Ding, Y., Tan, F., Qin, Z., Cao, M., Choo, K.K.R., Qin, Z.: DeepKeyGen: a deep learning-based stream cipher generator for medical image encryption and decryption. IEEE Trans. Neural Netw. Learn. Syst. **33**(9), 4915–4929 (2021)
28. Ahmad, I., Shin, S.: A perceptual encryption-based image communication system for deep learning-based tuberculosis diagnosis using healthcare cloud services. Electronics **11**(16), 2514 (2022)

DDPG-FL: A Reinforcement Learning Approach for Data Balancing in Federated Learning

Bei Ouyang, Jingyi Li, and Xu Chen(✉)

Sun Yat-sen University, Guangzhou 510330, China
{ouyb9,lijy573}@mail2.sysu.edu.cn, chenxu35@mail.sysu.edu.cn

Abstract. Federated learning (FL) is a novel distributed machine learning framework aimed at preserving privacy. A significant challenge in federated learning is the presence of non-independent and non-identically distributed (Non-IID) data among clients. Since local data is generated in diverse environments, the data distribution across data partitions may differ considerably, leading to slower and less accurate model training and increased communication overhead. In this paper, we propose a reinforcement learning-based federated learning data balancing algorithm on Non-IID data, DDPG-FL, which does not require the collection or inspection of any private information and does not introduce additional communication overhead. Additionally, DDPG-FL can be combined with existing FL algorithms as a data balancing plug-in. Experimental results on the MNIST, FMNIST, and CIFAR-10 datasets demonstrate that DDPG-FL enhances model performance while significantly reducing the number of communication rounds required for model convergence.

Keywords: Federated Learning · Non-Independent-and-Identically-Distributed Data · Reinforcement Learning

1 Introduction

Federated Learning (FL) [1–3] is an emerging machine learning framework that leverages privacy-sensitive data on mobile devices to perform data-driven analysis and decision-making tasks. As a distributed machine learning paradigm, FL allows a cluster of decentralized mobile devices at the edge to collaboratively train a shared machine learning model, while keeping all the raw training samples on the local devices. FL is demonstrated as a practical solution to mitigate the risk of privacy leakage and the dilemma of data silos.

An important challenge in FL is the non-independent and identically distributed data distribution (Non-IID) between clients [4]. Independent and Identically Distributed (IID) sampling of the training data is crucial to ensure that the stochastic gradient is an unbiased estimate of the full gradient. However, local data is often generated in different environments in practical scenarios, which can lead to significant differences in data distribution across data owners. For instance, patient data collected from different hospitals may differ significantly due to possible differences in patient demographics, disease type and severity, visit flow, and other factors. Severe data heterogeneity can easily lead to client drift [5], resulting in unstable convergence [6] and poor model performance [7,8]. Hence, conducting research on FL algorithms specifically designed for Non-IID data holds great significance.

In recent years, there are many research focusing on the non-iid issue in FL. Many efforts have been devoted to addressing client drift. FedProx [9] add a regularization term to limit the distance between the local model and the global model. Methods like SCAFFOLD [5] and MOON [10] use variance reduction techniques [11] or comparative learning to correct the client drift. Another direction is to directly rectify the cause of client drift, i.e., data heterogeneity. Zhao et al. [12] propose a data sharing method while obtaining a uniformly distributed global dataset is challenging. Therefore, some work [13–15] proposes data augmentation methods based on MixUp [16], Generative Adversarial Networks (GAN) [17] or Adversarial Learning. However, these methods expose the information of the original data, such as intermediate features and statistical information. To avoid additional information explosion, other methods solve data heterogeneity by balancing the data distribution of different clients. Astraea [18] creates the mediator to reschedule the training of clients according to local data distributions. However, direct access to raw data on individual clients is restricted due to privacy protection, making it impossible to analyze the data distribution on each client. Tang et al. [19] proposed an Imbalanced Weight Decay Sampling (IWDS) method that resamples data locally on the client side. But it can not dynamically adjust the data imbalance and cannot solve the problem of global data imbalance.

In this paper, we propose DDPG-FL, a federated learning algorithm tackling data heterogeneity based on deep reinforcement learning (DRL). DDPG-FL is designed to employ DRL to automatically learn sampling weights decision policies based on the observed client status and feedback rewards quantified by the model performance and data distributions. This algorithm aims to improve the efficiency and accuracy of FL by dynamically regulating the sampling weights of data on the client side to address data heterogeneity indirectly. Before local training, clients perform local data resampling to make the data distribution more similar, thus achieving data balancing. We propose using the Gini coefficient[1] as a metric to quantify data disparity in FL. And we formulate the process of determining the sampling weights of selected clients as a Markov decision process (MDP). Based on reinforcement learning, DDPG-FL dynamically regulates

---

[1] Gini Coefficient: a measure of the data disparity. Given a set of data $\{a_1, a_2, \ldots, a_N\}$, the Gini Coefficient is calculated as shown below, where $\mu = \frac{1}{N}\sum_{i=1}^{N} a_i$.

$$Gini = \frac{\sum_{i=1}^{N}\sum_{j=1}^{N}|a_i - a_j|}{2N^2\mu}$$

the sampling weights of the selected clients by considering each client's Gini coefficient and the global model's testing accuracy on each class. Meanwhile, DDPG-FL can be combined with existing FL algorithms as a data balancing plugin. The key contributions of this work can be summarized as follows:

- We propose using the Gini coefficient as a metric to quantify data disparity in FL and formulate the process of determining the resampling parameters of selected clients as a Markov decision process.
- We introduce DDPG-FL, a data balancing plug-in for FL based on deep reinforcement learning (DRL). This approach only requires clients to upload the Gini coefficient of the number of local categories, which follows the privacy-preserving principle and results in a negligible increase in communication overhead.
- We conduct extensive experiments on widely used datasets including MNIST, FMNIST, and CIFAR-10. We integrate DDPG-FL with popular FL algorithms, such as FedAvg, FedProx, FedNova, and SCAFFOLD. Experimental results demonstrate that DDPG-FL enhances model performance while significantly reducing the number of communication rounds required for model convergence.

2 Related Work

2.1 The Non-IID Challenge in Federated Learning

A key challenge in FL is the non-IID data distribution among different clients [4]. In many real-world scenarios, the data can be non-IID between clients. For example, various regions may have different disease distributions. Countries in the southern hemisphere may have a higher incidence of skin cancer patients than countries in the northern hemisphere because of the ozone hole.

Non-IID data distribution can significantly impact the accuracy and speed of convergence of FL algorithms. As the distribution of each local dataset is highly different from the global distribution, the local objective of each client is inconsistent with the global optima, leading to drift in the local updates [8]. During the local training stage, the local models are updated towards the local optima, which can be far from the global optima. This can cause the aggregated model to deviate from the global optima, particularly when local updates are large [5]. When using local updates to optimize the global model, the model can easily fall into a local optimum, leading to unstable training and slow convergence [20].

2.2 Federated Learning Algorithms on Non-IID Data Distribution

Extensive research has been conducted to address the challenge of non-IID data in FL from both the algorithmic and data perspective. In terms of algorithms, model regularization focuses on calibrating the optimization direction of the local model by adding regularization terms to limit the local model from being too far from the server model. FedProx [9] introduces an additional L_2 regularization term in the local objective function. SCAFFOLD [5] introduces variance between clients and applies variance reduction techniques [11] to correct client drift in its local updates. MOON [10] performs comparative learning between the server model and the client model.

There are also some algorithms that propose methods to correct client-side updates and accelerate convergence from an optimization perspective. FedNova [7] improves on FedAvg in the model aggregation phase by considering the case where different clients may perform different numbers of local steps. As the original cause of client drift is data heterogeneity, data-based approaches aim to address this problem by modifying the distribution. Some work focuses on some form of data sharing. Zhao et al. [12] propose a data sharing method where a small portion of data is shared among clients, resulting in a significant improvement in model accuracy. However, obtaining a uniformly distributed global dataset is challenging since the server does not know the data distribution among clients. Yoon et al. [13] addressed this issue by using the Mixup technique [16] for data augmentation, where clients can share mixed local data and collaborate to build a new global dataset. However, frequent data exchange may be subject to privacy attacks, so some studies propose data augmentation methods based on generative adversarial networks [17] and adversarial learning [21], like FedDPGAN [14], Fed-ZDAC [22]. All of the above methods can expose information about the original data, i.e., intermediate features, statistical information, and raw data to some extent, which violates the privacy protection principle of FL.

Other work addresses data heterogeneity from a data balancing perspective. Astraea [18] introduces mediators between server and clients. The mediator is assumed to have access to the local data distributions of the clients so that client rescheduling and data augmentation can be carried out accordingly. However, this approach requires uploading statistics on the amount of each type of data on the client, which may not be feasible in practice since in most cases, the client's data distribution is not accessible to the server. Tang et al. [19] proposed IWDS, in which the sampling weights of all data samples decay with training time. This approach allows clients to have more similar label sampling probabilities to converge faster in the early stage and to learn expertise better in the later stage. However, it is crucial to carefully adjust the hyperparameter β that control sampling weights and the decay rate ρ. Additionally, the method cannot solve the problem of global data imbalance. Therefore, a more dynamic and adaptive approach is necessary.

2.3 Deep Reinforcement Learning for Federated Learning

DRL [23] is an important unsupervised learning method that can automatically make decisions based on the policy learned by the agent interacting with the environment. Previous research has explored various applications of DRL techniques in the context of FL. Several studies [24–26] try to optimize client selection dynamically during runtime. PG-FFL [20] employs RL to learn a policy for assigning weights to model aggregation to improve fairness. Nguyen et al. [27] propose the use of Deep Q-Network (DQN) [23] for optimal decision-making regarding energy and channel allocation.

In recent years, actor-critic deep reinforcement learning methods, such as Deep Deterministic Policy Gradient (DDPG) [28], have obtained great achievements in solving optimization problems with large action spaces or continuous decision variables. In this paper, we formulate the process of determining the sampling weights of selected clients as a MDP problem and adopt DDPG to solve it. To the best of our knowledge, DDPG-FL is the first work that specifically employs DRL techniques to tackle the non-iid issue by performing data balancing in FL.

3 DDPG-FL

3.1 Formulation and Motivation

Assuming there are K clients with datasets $\mathcal{D}^{(1)}, \mathcal{D}^{(2)}, \cdots, \mathcal{D}^{(k)}$ respectively, and C is the total number of classes. We use T to denote the maximum number of communication rounds and c to denote the client selection ratio. At the beginning of the t-th round, $\lceil K \cdot c \rceil$ clients are randomly selected to form a subset of clients S. We use $V^{(k,i)}$ to represent the number of class i owned by client k, where $k \in \{1, 2, ... K\}$ and $i \in \{1, 2, ..., C\}$. The probability of client k sampling data from class i can be defined as $p^{(k,i)}$:

$$p^{(k,i)} = \frac{\left(V^{(k,i)}\right)^\beta}{\sum_{j=1}^{C}\left(V^{(k,j)}\right)^\beta},\qquad(1)$$

where β is the parameter that determines the sampling strategy. When β is set to be 1, Eq. 1 indicates the instance-balanced sampling, and when β is set to be 0, the possibility that every class to be selected is to $\frac{1}{C}$. By dynamically adjusting the value of β, we can regulate the relative sampling probabilities between classes. This allows us to make the class sampling probability between different clients become similar indirectly. Hence, it is essential to determine the appropriate parameter β for each selected client to ensure the effectiveness and efficiency of the FL training process.

In each round, the selected client k performs local data resampling on dataset $\mathcal{D}^{(k)}$. Next, clients k proceeds with E epochs of local training. Considering that direct access to raw data on individual clients is restricted due to privacy protection, we use the Gini coefficient to reflect the local distribution. In the t-th round, the local data distribution on client k is denoted by $d_t^{(k)} = \{V_t^{(k,1)}, \ldots, V_t^{(k,C)}\} \in \mathbb{R}^C$. The Gini coefficient of client k can be calculated as Eq. 2 defines.

$$g_t^{(k)} = \frac{\sum_{i=1}^{C}\sum_{j=1}^{C}\left|V_t^{(k,i)} - V_t^{(k,j)}\right|}{2C\sum_{i=1}^{C}V_t^{(k,i)}}.\qquad(2)$$

After t rounds, the accuracy of the global model ω_t on each class of samples in the test dataset is denoted as $acc_t = (acc_t^{(1)}, acc_t^{(2)}, \ldots, acc_t^{(C)})$. $r_t^{(a)}$ encourages a higher average accuracy while keeping the accuracy of each class as balanced as possible.

$$r_t^{(a)} = -\frac{\sum_{c=1}^{C} acc_t^{(c)}}{C} \log\left(\frac{\sum_{i=1}^{C} \sum_{j=1}^{C} \left|acc_t^{(i)} - acc_t^{(j)}\right|}{2C \sum_{i=1}^{C} acc_t^{(i)}}\right) \quad (3)$$

Our optimization problem can be defined as:

$$\max \sum_{k \in S} r_t^{(a)} \cdot \log(g_t^{(k)}) \quad (4)$$

In the absence of explicitly constructing estimates of accuracy of the global model, the problem is difficult to solve directly. Additionally, current decisions may have an impact on the future. Unlike the single-round optimization by using the model-based approach, DRL strives to learn a general action decision from past experience based on the current state and given reward. Therefore, we formulate the optimization problem as a Markov decision process (MDP) and then utilize DRL techniques.

3.2 Workflow

DPPG-FL can be regarded as an accessory plug-in to FL algorithms and can be combined with other FL algorithms to address the issue of non-iid. Algorithm 1 introduces DDPG-FedAvg as an example. Figure 1 provides an overview of the workflow. The DRL agent decides the resampling parameters $\beta^{(k)}$ for each selected client in each round. The clients then perform local data resampling according to the chosen parameters. The specific steps are as follows:

Step1: All available K clients check in server.

Step2: In the t-th round, $\lceil K \cdot c \rceil$ clients are randomly selected to form a subset of clients S. The DRL agent receives the current global model parameters ω_t and the model parameters $\omega_t^{(k)}$ of each selected client and combines them into a state vector $s_t = [\omega_t, \omega_t^{(k)} | k \in S]$. Based on the state, the DRL agent selects actions $a_t = \{\beta_t^{(k)} | k \in S\}$, where $\beta_t^{(k)}$ is the local resampling parameter of the selected client k, defined in Eq. 1.

Step3: Each selected client k initializes its model parameters using the global model by setting $\omega_t^{(k)} = \omega_t$. Next, the client k uses the resampling parameters $\beta_t^{(k)}$ to perform local data resampling. After E epochs of local training, the trained parameters are reported to the server ω_{t+1}^k. At the same time, selected clients send the local Gini coefficient $g_t^{(k)}$ to the server, which reflects the distribution of local data.

Step4: The server performs model aggregation (e.g. FedAvg) on the received parameters $\{\omega_{t+1}^{(k)} | k \in S\}$ to obtain the global model ω_{t+1}. Then, the server computes the reward r_t and the next state s_{t+1}.

Step5: The DRL agent puts the $\{s_t, a_t, r_t, s_{t+1}\}$ tuple into the experience replay buffer. The DRL agent samples data from the replay buffer and updates its parameters. Move into round $t + 1$ and repeat steps 2–5.

Repeat steps 2–5 until the end, e.g. to reach the target accuracy. Notably, in practical implementations, the DRL training process can be synchronized with the FL process without introducing additional computation time overhead due to the use of the experience replay buffer.

Algorithm 1. DDPG-FedAvg

Input: Number of communication round T, number of clients K, client selection ratio c, local epochs E, learning rate η;
Output: Final global model weight ω_T;
1: Initialization: global model with ω_0,
2: Randomly initialize critic network $Q(s, a|\theta^Q)$ and actor network $\mu(s|\theta^\mu)$ with weights θ^Q and θ^μ.
3: Initialize target network $\theta^{\mu^-} \leftarrow \theta^\mu, \theta^{Q^-} \leftarrow \theta^Q$
4: Initialize replay buffer R
5: **for** $t = 0, 1, \ldots, T - 1$ **do**
6:     Server randomly selects $\lceil K \times c \rceil$ clients to form the client subset S_t
7:     Compute the amount of data involved in training $n \leftarrow \sum_{k \in S_t} |\mathcal{D}^{(k)}|$
8:     Select action $a_t = \{\beta_t^{(k)} | k \in S\}$ according to the current policy
9:     The server sends the global model ω_t and a_t to the client subset S_t
10:    **for** each client $k \in S_t$ **do**
11:       $\omega_t^{(k)} \leftarrow \omega_t$
12:       Resampling on the local dataset $\mathcal{D}^{(k)}$ using the weights $\beta_t^{(k)}$
13:       Random gradient descent update to get $\omega_{t+1}^{(k)}$
14:       Send $\omega_{t+1}^{(k)}$ and $g_t^{(k)}$ to the server
15:    **end for**
16:    Update global model: $\omega_{t+1} \leftarrow \sum_{k \in S_t} \frac{|D^{(i)}|}{n} \omega_t^{(k)}$
17:    Compute the accuracy of the global model ω_{t+1} for each class of data on the
18:    test dataset
19:    Calculate r_t, s_{t+1}, store transition $\{s_t, a_t, r_t, s_{t+1}\}$ in R
20:    Sample a random minibatch of N transitions(s_i, a_i, r_i, s_{i+1}) from R
21:    Set $y_i = r_i + \gamma Q(s_i, \mu(s_{i+1}|\theta^{\mu^-})|\theta^{Q^-})$
22:    Update critic by minimizing the loss:
23:       $\mathcal{L} = \frac{1}{N} \sum_i (y_i - Q(s_i, a_i | \theta^Q))$
24:    Update the actor policy using the sampled policy gradient:
      $\nabla_{\theta^\mu} \mathcal{L} \approx \frac{1}{N} \sum_i \nabla_a Q(s, a | \theta^Q) \nabla_{\theta^\mu} \mu(s | \theta^\mu) | s_i$
25:    Update the target networks:
      $\theta^{\mu^-} = \tau \theta^\mu + (1 - \tau)\theta^{\mu^-}$
      $\theta^{Q^-} = \tau \theta^Q + (1 - \tau)\theta^{Q^-}$
26: **end for**
27: **return** ω_T

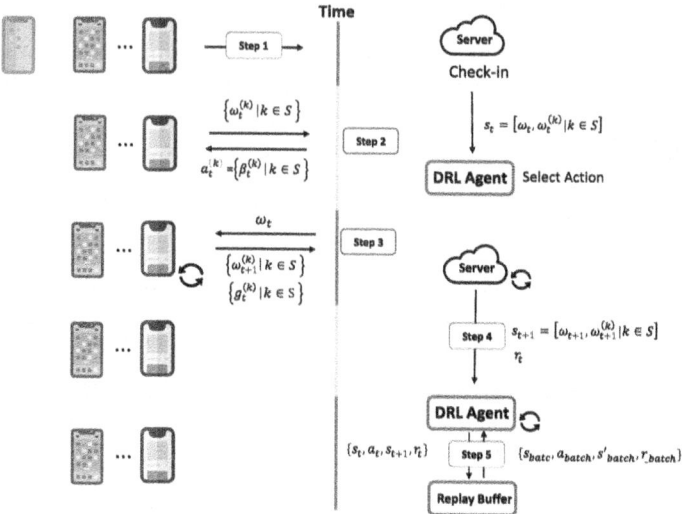

Fig. 1. The federated learning workflow with DDPG-FL

3.3 DRL Settings

This section models the local resampling parameter determination process as a MDP. The state is represented by the global model parameters and the model parameters for each client selected. Given the current state, the DRL agent computes the resampling parameters $\beta_t^{(k)}$ for each selected client k. The goal is to train the DRL agent so that the FL converges to the target accuracy as quickly as possible while guaranteeing performance. In addition, the DRL agent only needs to obtain the local model parameters and Gini coefficients from the clients, which do not require any private information to be collected and inspected, allowing for privacy preservation. The following is the description of the core components of reinforcement learning: state, action, and reward.

State: Wang et.al [24] observed an implicit connection between the distribution of training data on a device and the model weights trained based on those data through both empirical and mathematical analysis. In round t, the state is represented by a vector $s_t = [\omega_t, \omega_t^{(k)} | k \in S]$, where ω_t is the global model parameter after t rounds, and $\omega_t^{(k)}$ is the local model parameter of client k after t rounds.

Since CNN models can contain millions of weights, the resulting state space can be large. Using such a large state space to train RL algorithms is challenging. For this reason, we follow the approach proposed by Wang et al. [24] and use Principal Component Analysis (PCA) to reduce the model parameters. Specifically, we employ the sklearn.decomposition.PCA module.

Action: At the beginning of each round t, the agent determines the action $a_t = \{\beta_t^{(k)} | k \in S\}$, where $\beta_t^{(k)}$ is the local data resampling parameter of the selected client k defined in Eq. 1.

Reward: The reward r_t is defined by Eq. 5. The second term -1 encourages achieving the goal with fewer rounds. The goal of the DRL agent is to maximize the long-term cumulative return $R = \sum_{t=1}^{T} \gamma^{t-1} r_t$, where $\gamma \in (0, 1]$ is a discount factor.

$$r_t = \sum_{k \in S} r_t^{(a)} \cdot \log(g_t^{(k)}) - 1 \tag{5}$$

4 Experiment

4.1 Experimental Setup

Datasets and Models. We conducted experiments on three image datasets: MNIST [29], FMNIST [30], and CIFAR-10 [31]. The statistics of the datasets are summarized in Table 1. For these datasets, we used a CNN, which has two 5×5 convolution layers followed by 2×2 max pooling (the first with 6 channels and the second with 16 channels) and two fully connected layers with ReLU activation (the first with 120 units and the second with 84 units).

Table 1. The statistics of datasets in the experiments.

Datasets	training instances	test instances	features	classes
MNIST	60,000	10,000	784	10
FMNIST	60,000	10,000	784	10
CIFAR-10	50,000	10,000	1,024	10

Partition. Dirichlet distribution is an appropriate choice to simulate real-world data distribution. Hsu et al. [32] proposed Label-based Dirichlet Partition, which uses Dirichlet Sampling with a hyper-parameter α to simulate Label-based non-iid data distribution. This is a common practice in FL research to generate synthetic FL datasets [7,33]. We model non-iid data distributions using a Dirichlet distribution $Dir(\alpha)$, in which a smaller α indicates higher data heterogeneity, as it makes the distribution of $p_k(y)$ more biased for a user k. Figure 2 visualizes the distribution of MNIST dataset corresponding to different α when the number of clients is 10. α gradually increases from 0.1, 0.5, and 1.0, demonstrating the process of data distribution gradually becoming more uniform.

Baselines. We conducted FedAvg and recent effective FL algorithms that are proposed to address the non-iid problem, including FedProx, FedNova, and SCAFFOLD.

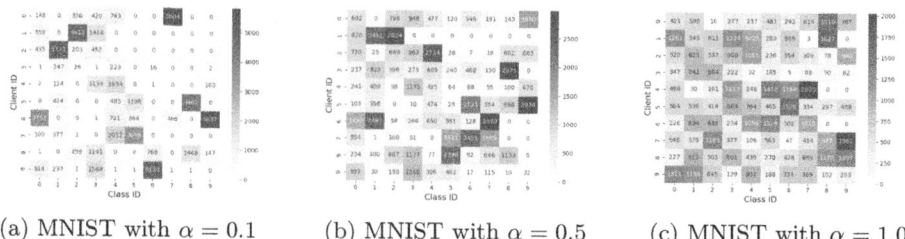

(a) MNIST with $\alpha = 0.1$ (b) MNIST with $\alpha = 0.5$ (c) MNIST with $\alpha = 1.0$

Fig. 2. Visualization of MNIST dataset on 10 clients with non-iid partition. The color bar denotes the number of data samples. Each square represents the number of data samples of a specific class in a client.

Performance metrics. In FL, reducing the number of communication rounds is crucially important due to the limited computation capacity and network bandwidth of mobile devices. Therefore, the performance metrics used in this paper are the model accuracy and the number of communication rounds required for the model to achieve the target accuracy.

4.2 Experiment Results

Comparison with IWDS. In our experiments, we set the number of clients, K, to 10. The local training epoch E is 1, and the client selection ratio, c, is set to 1.0. The maximum number of communication rounds T is 150 for the MNIST and FMNIST datasets and 250 for the CIFAR-10 dataset. As for IWDS, $\beta_0 = 0.9999$, $\beta_m = 0.95$, decay rate $\rho = 0.992$.

Table 2. Comparison of accuracy and communication rounds for baseline algorithms using DDPG-FL and IWDS under $Dir(0.5)$ data partitioning.

	Accuracy (%)			Communication Rounds		
	MNIST	FMNIST	CIFAR-10	MNIST	FMNIST	CIFAR-10
FedAvg	**99.07**	89.29	64.66	70	70	101
FedAvg+IWDS	99.01	89.14	65.07	**49**	45	100
FedAvg+DDPG-FL	99.06	**89.33**	**65.14**	63	**42**	**87**
FedProx	99.01	88.99	64.28	70	78	123
FedProx+IWDS	**99.09**	89.14	64.31	66	71	123
FedProx+DDPG-FL	99.08	**89.99**	**65.34**	**54**	**43**	**85**
FedNova	99.06	89.06	63.99	80	70	123
FedNova+IWDS	99.00	89.24	65.01	82	66	115
FedNova+DDPG-FL	**99.09**	**89.32**	**65.84**	**69**	**45**	**88**
SCAFFOLD	99.15	**89.53**	65.23	48	67	106
SCAFFOLD+IWDS	99.17	89.19	63.53	45	61	121
SCAFFOLD+DDPG-FL	**99.20**	89.46	**66.83**	**22**	**38**	**64**

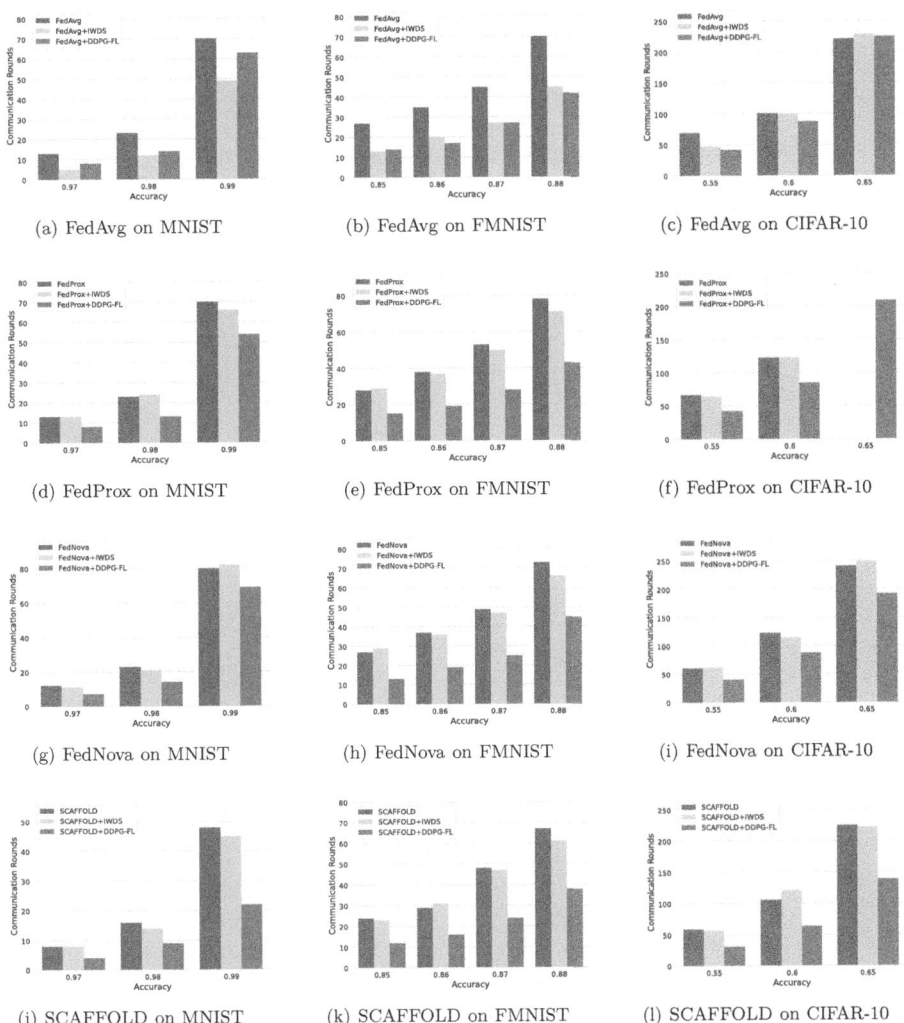

Fig. 3. From left to right are the experimental results on the MNIST, FMNIST, CIFAR-10 datasets, respectively. The horizontal coordinate is the test accuracy and the vertical coordinate is the number of communication rounds. Each subplot compares the number of rounds required by the global model to reach a specific accuracy on the test set when using different data balancing algorithms. The missing positions indicate that the corresponding accuracy was not reached until the maximum number of communication rounds T.

The final model accuracy of each algorithm on different datasets is recorded in Table 2. In most cases, DDPG-FL achieves optimal performance, especially on the CIFAR-10 dataset. Even when it falls slightly short of the optimal performance, DDPG-FL achieves accuracy levels within a very close range. Entrise in

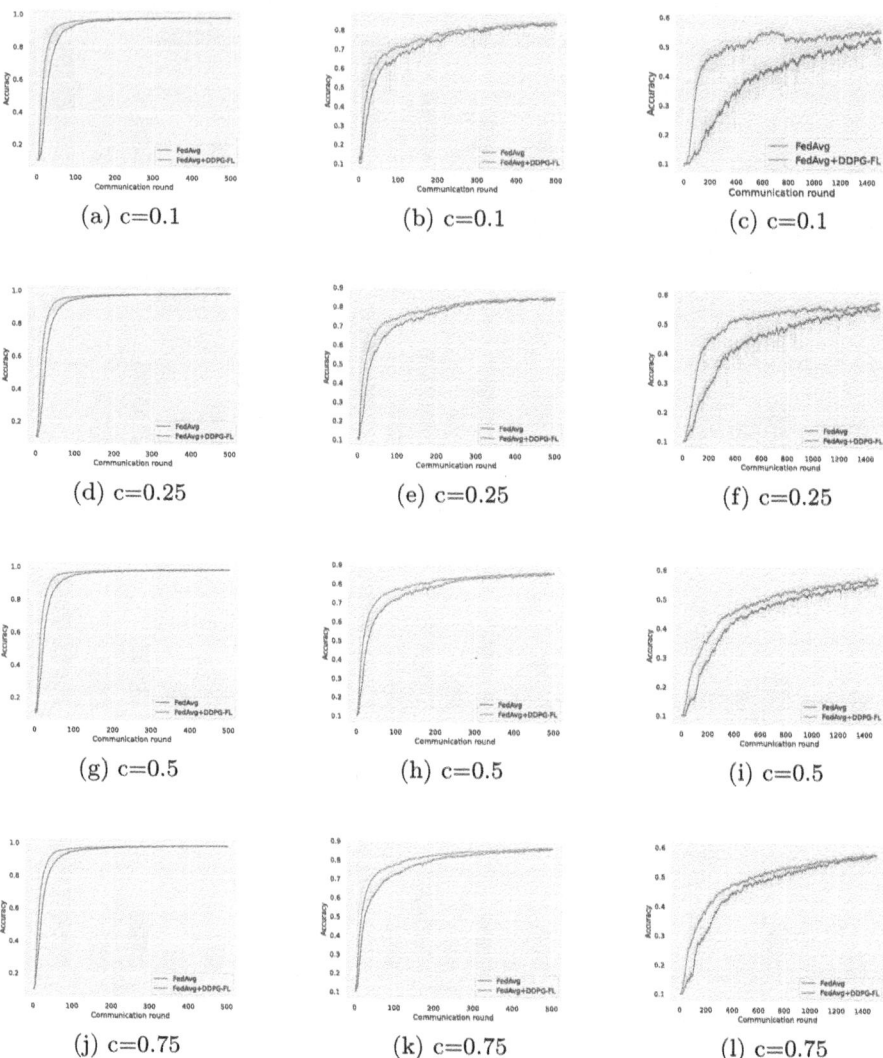

Fig. 4. FedAvg training with different levels of parallelism. From left to right are the experimental results on the MNIST, FMNIST and CIFAR-10 datasets, respectively.

Table 2 shows the number of communication rounds required to achieve specific test set accuracy: 99% for the CNN on MNIST, 88% for FMNIST, and 60% for CIFAR-10. The results demonstrate that FL algorithms incorporating DDPG-FL are highly effective in reducing the number of rounds required to achieve the desired accuracy levels. DDPG-FL outperforms IWDS in all cases except for the FedAvg algorithm on the MNIST dataset.

Figure 3 illustrates the convergence of the FL algorithms using different data balancing algorithms on $Dir(0.5)$ data distribution. DDPG-FL can effectively accelerate the model convergence speed and reduce communication overhead. Combining the above experimental results, DDPG-FL can be used as a data balancing plug-in for FL algorithms FedAvg, FedProx, FedNova, and SCAFFOLD to improve model performance while greatly reducing the number of communication rounds required for model convergence.

Increasing Parallelism. We also conducted experiments to evaluate the performance of DDPG-FL with different client participation ratios. The number of clients, K, was set to 100, and we varied the client participation ratio from 0.1, 0.25, 0.5, to 0.75 to examine the impact of different levels of parallelism in FL. For the MNIST and FMNIST datasets, the maximum number of communication rounds was set to 500, while for the CIFAR-10 dataset, it was set to 1500. It can be observed from Fig. 4 that the improvement achieved by adding DDPG-FL plug-in is more significant when the client participation ratio is low.

5 Conclusion

The non-iid data distribution among clients poses significant challenges in federated learning. In this paper, we use Gini coefficient to measure the data disparity and formulate the process of determining the sampling weights of selected clients as a Markov decision process to perform data balancing. We present a federated learning data balancing algorithm, DDPG-FL, based on reinforcement learning. DDPG-FL dynamically regulates the sampling weights of the selected clients during the FL training process. Additionally, DDPG-FL can be combined with existing FL algorithms as a data balancing plug-in. Through experiments on MNIST, FMNIST, and CIFAR-10 datasets, DDPG-FL can be used as a data balancing plug-in for the FedAvg, FedProx, FedNova, and SCAFFOLD to improve the model performance while greatly decrease the number of communication rounds required for model convergence.

References

1. McMahan, B., Moore, E., Ramage, D., Hampson, S., y Arcas, B.A.: Communication-efficient learning of deep networks from decentralized data. In: Artificial Intelligence and Statistics, pp. 1273–1282. PMLR (2017)
2. Kairouz, P., et al.: Advances and open problems in federated learning. Foundations and Trends® in Machine Learning **14**(1–2), 1–210 (2021)
3. Park, J., Samarakoon, S., Bennis, M., Debbah, M.: Wireless network intelligence at the edge. Proc. IEEE **107**(11), 2204–2239 (2019)
4. Kaissis, G.A., Makowski, M.R., Rückert, D., Braren, R.F.: Secure, privacy-preserving and federated machine learning in medical imaging. Nature Mach. Intell. **2**(6), 305–311 (2020)
5. Karimireddy, S.P., Kale, S., Mohri, M., Reddi, S., Stich, S., Suresh, A.T.: Scaffold: Stochastic controlled averaging for federated learning. In: International Conference on Machine Learning, pp. 5132–5143. PMLR (2020)

6. Li, X., Huang, K., Yang, W., Wang, S., Zhang, Z.: On the convergence of fedavg on non-iid data. In: International Conference on Learning Representations (2020)
7. Wang, J., Liu, Q., Liang, H., Joshi, G., Poor, H.V.: Tackling the objective inconsistency problem in heterogeneous federated optimization. Adv. Neural. Inf. Process. Syst. **33**, 7611–7623 (2020)
8. Zhao, Y., Li, M., Lai, L., Suda, N., Civin, D., Chandra, V.: Federated learning with non-iid data (2018)
9. Li, T., Sahu, A.K., Zaheer, M., Sanjabi, M., Talwalkar, A., Smith, V.: Federated optimization in heterogeneous networks. Proc. Mach. Learn. Syst. **2**, 429–450 (2020)
10. Li, Q., He, B., Song, D.: Model-contrastive federated learning. In: 2021 IEEE/CVF Conference on Computer Vision and Pattern Recognition (CVPR), pp. 10708–10717. IEEE Computer Society (2021)
11. Johnson, R., Zhang, T.: Accelerating stochastic gradient descent using predictive variance reduction. In: Advances in Neural Information Processing Systems, vol. 26 (2013)
12. Shoham, N., et al.: Overcoming forgetting in federated learning on non-iid data (2019)
13. Yoon, T., Shin, S., Hwang, S.J., Yang, E.: Fedmix: Approximation of mixup under mean augmented federated learning. In: International Conference on Learning Representations (2021)
14. Zhang, L., Shen, B., Barnawi, A., Xi, S., Kumar, N., Wu, Y.: Feddpgan: federated differentially private generative adversarial networks framework for the detection of Covid-19 pneumonia. Inf. Syst. Front. **23**(6), 1403–1415 (2021)
15. Goetz, J., Tewari, A.: Federated learning via synthetic data (2020)
16. Zhang, H., Cisse, M., Dauphin, Y.N., Lopez-Paz, D.: mixup: Beyond empirical risk minimization. In: International Conference on Learning Representations (2018)
17. Goodfellow, I., et al.: Generative adversarial networks. Commun. ACM **63**(11), 139–144 (2020)
18. Duan, M., et al.: Astraea: Self-balancing federated learning for improving classification accuracy of mobile deep learning applications. In: 2019 IEEE 37th International Conference on Computer Design (ICCD), pp. 246–254. IEEE (2019)
19. Tang, Z., et al.: Data resampling for federated learning with non-iid labels. In: Proceedings of the International Workshop on Federated and Transfer Learning for Data Sparsity and Confidentiality in Conjunction with IJCAI, vol. 2021 (2021)
20. Sun, Y., Si, S., Wang, J., Dong, Y., Zhu, Z., Xiao, J.: A fair federated learning framework with reinforcement learning. In: 2022 International Joint Conference on Neural Networks (IJCNN), pp. 1–8. IEEE (2022)
21. Huang, L., Joseph, A.D., Nelson, B., Rubinstein, B.I., Tygar, J.D.: Adversarial machine learning. In: Proceedings of the 4th ACM Workshop On Security and Artificial Intelligence, pp. 43–58 (2011)
22. Hao, W., et al.: Towards fair federated learning with zero-shot data augmentation. In: Proceedings of the IEEE/CVF Conference on Computer Vision and Pattern Recognition, pp. 3310–3319 (2021)
23. Mnih, V., et al.: Playing atari with deep reinforcement learning (2013)
24. Wang, H., Kaplan, Z., Niu, D., Li, B.: Optimizing federated learning on non-iid data with reinforcement learning. In: IEEE INFOCOM 2020-IEEE Conference on Computer Communications, pp. 1698–1707. IEEE (2020)
25. Zhang, S.Q., Lin, J., Zhang, Q.: A multi-agent reinforcement learning approach for efficient client selection in federated learning. In: Proceedings of the AAAI Conference on Artificial Intelligence. vol. 36, pp. 9091–9099 (2022)

26. Kim, Y.G., Wu, C.J.: Autofl: Enabling heterogeneity-aware energy efficient federated learning. In: MICRO-54: 54th Annual IEEE/ACM International Symposium on Microarchitecture, pp. 183–198 (2021)
27. Nguyen, H.T., Luong, N.C., Zhao, J., Yuen, C., Niyato, D.: Resource allocation in mobility-aware federated learning networks: a deep reinforcement learning approach. In: 2020 IEEE 6th World Forum on Internet of Things (WF-IoT), pp. 1–6. IEEE (2020)
28. Lillicrap, T.P., et al.: Continuous control with deep reinforcement learning (2015)
29. LeCun, Y., Bottou, L., Bengio, Y., Haffner, P.: Gradient-based learning applied to document recognition. Proc. IEEE **86**(11), 2278–2324 (1998)
30. Xiao, H., Rasul, K., Vollgraf, R.: Fashion-mnist: a novel image dataset for benchmarking machine learning algorithms (2017)
31. Krizhevsky, A., Hinton, G., et al.: Learning multiple layers of features from tiny images (2009)
32. Hsu, T.M.H., Qi, H., Brown, M.: Measuring the effects of non-identical data distribution for federated visual classification (2019)
33. Reddi, S.J., et al.: Adaptive federated optimization. In: International Conference on Learning Representations (2020)

Adaptive Recovery with Reinforcement Learning in Cloud-of-Clouds Storage Systems

Jiajie Shen, Bochun Wu[✉], Wang Xiang, Zeyu Zhao, and Kai Zhang

Informatization Office, Fudan University, Shanghai 200433, China
{jiajieshen,wubochun,xiangw,fdzzy,zhangkai}@fudan.edu.cn

Abstract. Cloud-of-clouds storage systems are widely used in online applications, where user data are encrypted, encoded, and stored in multiple clouds. When some cloud nodes fail, the storage systems need to reconstruct the lost data and store it in the substitute nodes. It is a challenge to reduce the time of data recovery process to ensure the data reliability. In this paper, we adopt a reinforcement learning-based data recovery (RLDR) approach to reduce the regeneration time. By employing Mento-Carlo method, our approach can construct the tree-topology based regeneration process, (*a.k.a.* regeneration tree), to effectively reduce the regeneration time. Through theoretical analysis, we apply the information flow graph to optimize the network traffic between clouds for given regeneration tree. We conduct extensive experiments with real-world traces to verify the merit of RLDR. The experimental results demonstrate that our scheme can significantly accelerate regeneration process. Specifically, RLDR can reduce the regeneration time by up to 92% and improve the throughput by up to 1210%, compared with the state-of-the-art alternatives. To the best of our knowledge, this is the first work which adopts the reinforcement learning paradigm to reduce the regeneration time in cloud-of-clouds storage systems.

Keywords: Data recovery · Reinforcement learning · Cloud-of-clouds storage · Network coding · Performance optimization

1 Introduction

Cloud-of-clouds storage systems [1–4] are widely used in current online applications to store user data. It uses a dispersal algorithm [5] to encrypt the user data to several data shares, encode the data shares to generate parity shares, and store these shares to different cloud. By the encryption framework (*e.g.*, AONT [6]), the attackers cannot infer (even patrial) the original use data, when some clouds are compromised. The user device can also download the data from the clouds with high network bandwidth to enhance the efficiency of read operations [4].

However, such a dependable scheme does not widely used in current online applications, since it typically suffers from high data recovery overhead [7]. When some cloud nodes fail, the storage systems need to reconstruct lost data in substitute cloud nodes (*a.k.a.* newcomers) as soon as possible to ensure the data reliability [8]. Since the newcomers need to download the data from other clouds (*a.k.a.* providers), the network condition between clouds severely affects the efficiency of regeneration process [9]. To this end, state-of-the-art data recovery schemes typically adopt a heuristic approach to construct the regeneration tree.

However, it is difficult to improve the performance of regeneration process. First, generating regeneration tree can be transformed to a well-known NP-hard problem [9], which is typically quite time consuming. Second, the storage systems need to quickly recover the lost data to maintain data reliability. In other words, the regeneration scheme should construct the regeneration tree within a time limit.

To solve the above practical challenges, we propose an reinforcement learning-based data recovery (RLDR) scheme. By analyzing the performance of cloud storage systems, we notice that the performance of regeneration process is severely affected by the network bandwidth between clouds. Accordingly, we apply reinforcement learning-based paradigms to construct regeneration tree and determine the network traffic. The experimental results show that RLDR can significantly improve the performance of regeneration process. Specifically, we make the following contributions.

- **Adopting reinforcement learning to construct a regeneration tree.** We formulate the problem of constructing a regeneration tree as a combination search process, and apply a reinforcement learning-based scheme to perform the search operations. To the best of our knowledge, this is the first work to adopt the reinforcement learning-based paradigm to improve the performance of regeneration process.
- **Optimizing traffic according to network topology.** By employing the information traffic graph, we obtain the bound of network traffic for cooperative regeneration process. To speed up the search process, we propose an efficient method to directly calculate regeneration time for a given network topology. Accordingly, RLDR can quickly estimate the regeneration time without a complex calculation process.
- **System implement.** To verify the merit of RLDR, we implement various data recovery schemes, and evaluate the performance of these approaches with real-world traces. We further set the network bandwidth between storage nodes to simulate the network condition between clouds in a local cluster. According to the experimental results, our scheme can reduce the regeneration time by up to 92% and improve the throughput by up to 1210% compared with the state-of-the-art alternatives.

2 Background and Motivation

2.1 Preliminary of Cloud-of-Clouds Storage

Cloud-of-clouds storage is widely adopted to store user data (*a.k.a.* secret message) in online applications. When cloud nodes fail, the storage systems need to employ new cloud nodes (*a.k.a.* newcomers) and download the data from survive clouds (*a.k.a.* providers) to recover the lost data [8]. All lost data should be recovered as soon as possible, so that storage systems can ensure the reliability of user data [9].

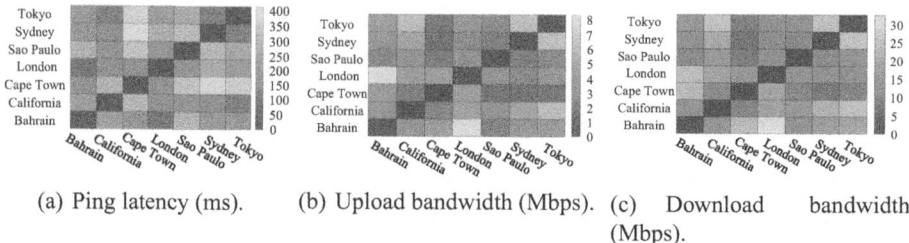

(a) Ping latency (ms). (b) Upload bandwidth (Mbps). (c) Download bandwidth (Mbps).

Fig. 1. The transmission performance of virtual machines between different data centers in Amazon EC2. We present the ping latency and bandwidth of both upload and download operations by using the Speedtest toolkit [10].

To solve the above problems, many studies propose various data recovery schemes to reduce the time of regeneration process, *e.g.*, Regenerating Code (RC) [8], Cooperative Regenerating Code (CRC) [11] RCTREE [12], CTREE [13], Heterogeneity-aware Cooperative Regeneration (HCR) [14], Reconstruction Codes (LRC) [15], NCCloud [7], Flexible Tree-structured Regeneration (FTR) [9], Optimal Multiple Failures Repair (OMFR) [16], PUSH [17], Flexible Regenerating Code (FRC) [18], and Flexible Cooperative Regenerating Code (FCRC) [19].

2.2 Measurement Setup

We use 7 Virtual Machines (VMs) in Amazon EC2 [20], as deployed in the data centers in Tokyo, Sydney, San Paulo, London, Cape Town, California, and Bahrain. During data recovery process, clouds need to perform the computation operations and to transfer data with each other. Since the transmission operations, rather than computation operations, typically suffer from low throughput [5], we focus on evaluating the performance of transmission operations.

We deploy Speedtest toolkit [10] in all the VMs to evaluate the performance of transmission operations between different VMs. To identify the bottleneck of regeneration process, we measure the ping latency and bandwidth for upload and download operations between clouds. The bandwidth is calculated as the results of total transferred data volume divided by the transmission latency, which can be collected by the Speedtest toolkit. The measurement results are shown in Fig. 1.

2.3 Measurement Results

According to measurement results, we can obtain the following key observations.

- **High transmission overhead.** Since clouds need to transfer data through Internet, they typically suffer from high transmission latency and low network bandwidth. The regeneration scheme needs to fully utilize the network bandwidth to reduce the time of recovery operations.

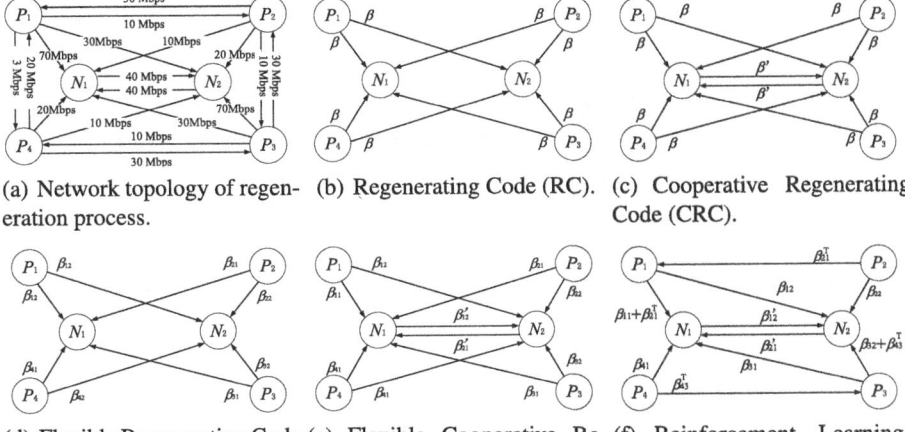

(a) Network topology of regeneration process.
(b) Regenerating Code (RC).
(c) Cooperative Regenerating Code (CRC).
(d) Flexible Regenerating Code (FRC).
(e) Flexible Cooperative Regenerating Code (FCRC).
(f) Reinforcement Learning-based Data Recovery (RLDR).

Fig. 2. Example of information flow graph with $d = 4$ providers and $r = 2$ newcomers. Regeneration schemes should determine the regeneration tree and network traffic between multiple cloud storage nodes.

- **Heterogeneous network bandwidth.** Since clouds are geographically located, the transmission latency and network bandwidth are typically heterogeneous. In Fig. 1, the maximum latency is about 5 times as its minimum value, 8 times for network bandwidth of upload operations, and 7 times for download operations.

2.4 Beyond the Regeneration Process: A Motivation Example

As illustrated in Fig. 2, we consider the example of the regeneration process with $d = 4$ providers and $r = 2$ newcomers. We assume that the redundant data is coded as $(n = 6, d = 4, k = 2, r = 2)$-MDS code, where any $k = 2$ out of $n = 6$ cloud nodes can reconstruct the original file. Suppose the size of original file as $M = 480$ Mb, and each storage node stores $\alpha = \frac{M}{k} = 240$ Mb of data. The network topology with heterogeneous link capacity is shown in Fig. 2(a).

To recover the lost data in newcomers N_1 and N_2, Regenerating Code (RC) requires a newcomer to download $\beta = \frac{M}{k(d-k+1)} = 80$ Mb from every provider. The regeneration process of RC is shown in Fig. 2(b). Since RC transfers same data volume on each link, the regeneration time is severely affected by the bottleneck bandwidth c_{bot}. In this example, the regeneration time of RC is $t_{\text{RC}} = \frac{\beta}{c_{\text{bot}}} = \frac{80\text{Mb}}{10\text{Mbps}} = 8$ seconds.

To reduce the network traffic between clouds, Cooperative Regenerating Code (CRC) allows the newcomers exchange data during the regeneration process. The regeneration process of CRC is shown in Fig. 2(C). The network traffic of CRC for each link is $\beta = \beta' = \frac{M}{k(d-k+r)} = 60$ Mb. In this example, the regeneration time of CRC is $t_{\text{CRC}} = \max\{\frac{60\text{Mb}}{10\text{Mbps}}, \frac{60\text{Mb}}{30\text{Mbps}}\} = 6$ seconds. Although CRC can reduce the network traffic, it cannot significantly reduce the regeneration time.

To reduce the regeneration time of RC and CRC, Flexible Regenerating Code (FRC) transfers the various network traffic between clouds. Regeneration process of FRC is shown in Fig. 2(d). The network traffic of newcomer N_1 is $\{\beta_{1,1}, \beta_{2,1}, \beta_{3,1}, \beta_{4,1}\} = \{120\text{Mb}, 40\text{Mb}, 120\text{Mb}, 80\text{Mb}\}$, and that of newcomer N_2 is $\{\beta_{1,2}, \beta_{2,2}, \beta_{3,2}, \beta_{4,2}\} = \{120\text{Mb}, 80\text{Mb}, 120\text{Mb}, 40\text{Mb}\}$. Therefore, the regeneration time of FRC is

$$t_{\text{FRC}} = \max\{\frac{40\text{Mb}}{10\text{Mbps}}, \frac{80\text{Mb}}{20\text{Mbps}}, \frac{120\text{Mb}}{30\text{Mbps}}, \frac{120\text{Mb}}{70\text{Mbps}}\} = 4 \text{ seconds.} \qquad (1)$$

Similarly, Flexible Cooperative Regenerating Code (FCRC) transfers flexible network traffic for cooperative regeneration process in Fig. 2(e). In this example, the regeneration time of FCRC is

$$t_{\text{FCRC}} = \max\{\frac{24\text{Mb}}{10\text{Mbps}}, \frac{48\text{Mb}}{20\text{Mbps}}, \frac{72\text{Mb}}{30\text{Mbps}}, \frac{96\text{Mb}}{40\text{Mbps}}\} = 2.4 \text{ seconds.} \qquad (2)$$

The performance of FRC and FCRC is still severely affected by the bottleneck links between clouds. To avoid the bottleneck link in regeneration tree, RLDR can generate regeneration tree for cooperative regeneration process. By observing the network topology in this example, we notice that $c(P_2, N_1)$ and $c(P_4, N_2)$ are performance bottleneck of regeneration process. RLDR can bypass these bottleneck links to speedup the regeneration process. Figure 2(f) shows the regeneration process of RLDR. The regeneration time of RLDR is

$$t_{\text{RLDR}} = \max\{\frac{40\text{Mb}}{20\text{Mbps}}, \frac{60\text{Mb}}{30\text{Mbps}}, \frac{80\text{Mb}}{40\text{Mbps}}, \frac{140\text{Mb}}{70\text{Mbps}}\} = 2 \text{ seconds.} \qquad (3)$$

3 System Modeling and Problem Formulation

3.1 Network Topology

Assume a file consisting of M bytes of data is encoded to generate $n * \alpha$ shares, which are stored in n cloud nodes. Each node holds α bytes of data. The user can reconstruct the file by accessing to k nodes. When r cloud nodes fail, a newcomer can access d cloud nodes to generate r shares.

To ensure the newcomers can recover the lost data, the providers need to transfer enough data volume to the newcomers. For example, when the storage systems use CRC to perform the regeneration process, the minimum data volume for the Minimum Storage Cooperative Regeneration (MSCR) point can be calculated by the following expression:

$$\alpha = \frac{M}{k}, \beta = \beta' = \frac{M}{k*(d-k+r)}. \qquad (4)$$

We use complete graph $G(V, E)$ to represent this overlay network topology, which consists of d providers $\{P_1, P_2, ..., P_d\}$ and r newcomers $\{N_1, N_2, ..., N_r\}$.

For any two nodes n_i and n_j, let $f(C_i, C_j)$ and $c(C_i, C_j)$ denote the data volume transferred and the link capacity from cloud C_i to cloud C_j. Since cloud nodes can transfer the data simultaneously, we can define the regeneration time D as

$$D = \max\{\frac{f(C_i, C_j)}{c(C_i, C_j)} | (C_i, C_j) \in E\}. \qquad (5)$$

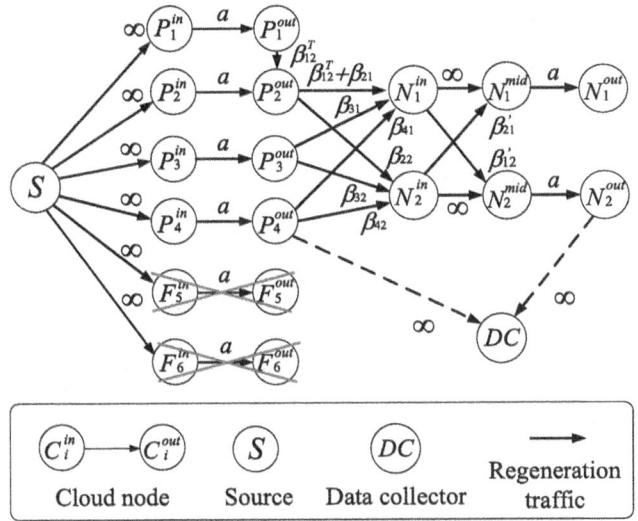

Fig. 3. Example of information flow graph for the tree-topology based cooperative regeneration process.

According to Eq. (5), the cloud-of-clouds storage systems minimize the regeneration time, while ensure the newcomers can receive the sufficient data from the providers during the regeneration process to reconstruct lost data.

3.2 Network Traffic for Cooperative Regeneration Tree

Let $\beta_{i,j}^T$ represent the data volume transferred from provider P_i to provider P_j, $\beta_{i,j}$ represent the data volume required to transfer from provider P_j to newcomer N_j to perform the regeneration process, and $\beta'_{i,j}$ represent the number of blocks transferred from newcomer N_i to newcomer N_j. To maintain the MDS property of regeneration process with tree-topology, the providers need to transmit sufficient data to the newcomers. We employ the information flow graph to obtain the network traffic between clouds and minimize the regeneration time. For a given regeneration process, we define total network traffic from provider P_i to provider P_j as

$$\beta_{i,j}^T = \sum \beta_{x,y} | \beta_{x,y} \in S_p(P_i, P_j), \tag{6}$$

where $S_p(P_i, P_j)$ is the flow set on link from provider P_i to provider P_j.

We show an example of information flow graph in Fig. 3. In this example, network traffic from provider P_1 to provider P_2 is $\beta_{1,2}^T = \beta_{1,1} + \beta_{1,2}$. Let β_l represent the network traffic between providers and newcomers (*i.e.*, $\beta_{i,j}$ and $\beta'_{i,j}$). For a given regeneration tree T, suppose network traffic $\beta_1, \beta_2, ..., \beta_e$ to satisfy

$$\beta_1 \leq \beta_2 \leq ... \leq \beta_e, \tag{7}$$

where parameter $e = d * r + r * (r - 1)$.

3.3 Network Traffic Optimization

We order n cloud nodes as $\{P_1, P_2, ..., P_d, N_1, N_2, ..., N_r\}$, and calculate the network traffic for a given link from cloud C_i to cloud C_j by the following expression.

$$f(C_i, C_j) = \begin{cases} \beta_{i,j}^T, & C_i \in S_P, C_j \in S_P \\ \sum_{0 < x \leq d} \beta_{x,i}^T + \beta_{i,m}, & C_i \in S_P, C_j \in S_N \\ \beta'_{l,m}, & C_i \in S_N, C_j \in S_N \end{cases}, \quad (8)$$

where S_P is the provider set, S_N is the newcomer set, parameter $l = i - d$, and parameter $m = j - d$.

According to Eqs. (5), (6) and (8), we can optimize the regeneration time by

$$\min \quad D = \max\{\frac{f(C_i, C_j)}{c(C_i, C_j)} | (C_i, C_j) \in E\} \quad (9a)$$

$$s.t. \sum_{l=1}^{d+r-k+j} \beta_l \geq \min\{(d-k+j)\beta, \alpha\}, \quad (9b)$$

$$0 \leq \beta_1 \leq \beta_2 \leq ... \leq \beta_d \leq \alpha. \quad (9c)$$

where T represents the regeneration time defined in Eq. (5).

According to Eq. (9), we can determine the network traffic by the Linear Programming (LP) [9]. Since the regeneration time is severely affected by the network bandwidth between clouds, the storage systems need to construct the regeneration tree to minimize the regeneration time.

3.4 Expectation of Regeneration Time

Suppose there are g regeneration strategies that can recover the lost data in newcomers. Let $D_y(C_i, C_j)$ represent the regeneration time of the y-th regeneration strategy whose process includes link from cloud C_i to cloud C_j. Accordingly, we define the expect regeneration time for given link for cloud C_i to cloud C_j as

$$l(C_i, C_j) = \overline{D_y(C_i, C_j)}, \quad 0 < y \leq g, \quad (10)$$

where $\overline{X}$ represents the average function for set X.

According to Eq. (10), we use matrix $\mathbf{L}$ to represent the expect regeneration time for all the links in given network topology $G(V, E)$ as

$$\mathbf{L} = \begin{bmatrix} 0 & l(C_1, C_2) & \cdots & l(C_1, C_n) \\ l(C_2, C_1) & 0 & \cdots & l(C_2, C_n) \\ \vdots & \vdots & \ddots & \vdots \\ l(C_n, C_1) & l(C_n, C_2) & \cdots & 0 \end{bmatrix}. \quad (11)$$

According to Eq. (11), we propose our regeneration scheme, namely RLDR, which tends to select the links with low expect regeneration time to construct the regeneration tree, so that the storage systems can fully utilize the network bandwidth between clouds to improve the performance of regeneration process.

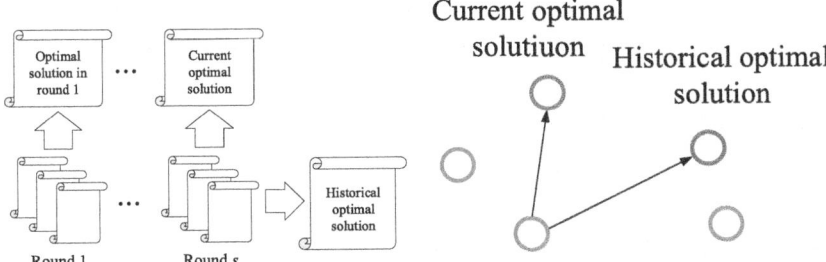

(a) Search process of optimal solutions.   (b) Updating the expect regeneration time.

Fig. 4. Reinforcement learning-based search process for the regeneration strategy.

4 Proposed RLDR Scheme

4.1 Generating Regeneration Tree

Since there are r newcomers each of which needs to receive data from d providers, we apply two vectors to record the regeneration tree for r newcomers as

$$R_i = \{W_1, W_2, ..., W_r\}, W_i = \{Z_1, Z_2, ..., Z_d\}, \qquad (12)$$

where W_i represents the regeneration tree that transfer data to newcomer N_i, and Z_i represents the parent node for newcomer P_i in regeneration tree.

We can easily identify whether the regeneration tree has been searched according to Eq. (12). To quickly select the links to generate the regeneration tree with low regeneration time, we select the links with small expect regeneration time to generate regeneration tree according to the expect regeneration time in Eq. (10). Accordingly, we define the link value

$$v(C_i, C_j) = l_{\max} - l(C_i, C_j), \text{ for } l_{\max} = \max(\mathbf{L}), \qquad (13)$$

where $\max(\mathbf{L})$ is the function that calculates the maximum value in matrix $\mathbf{L}$.

Since the newcomers exchange data during cooperative regeneration process, we only need to determine how to transfer data from providers to newcomers. For newcomer N_i, we employ Prim algorithm [9] to generate regeneration tree.

We start with a regeneration tree containing only the newcomer N_i, and iteratively add the providers to the tree until it spans all providers [9]. During constructing the regeneration tree, RLDR tends to select links with small expect regeneration time (*i.e.*, high link value). Therefore, we define the possibility to select given link (C_i, C_j) as

$$p(C_i, C_j) = \begin{cases} \dfrac{v(C_i, C_j)}{\sum_{C_y \in S_m} \sum_{x=1}^{n} v(C_x, C_y)}, & C_i \in S_P \text{ and } C_j \in S_m \\ 0, & \text{otherwise} \end{cases} \qquad (14)$$

where S_m is cloud set of the regeneration tree, when performing the Prim algorithm.

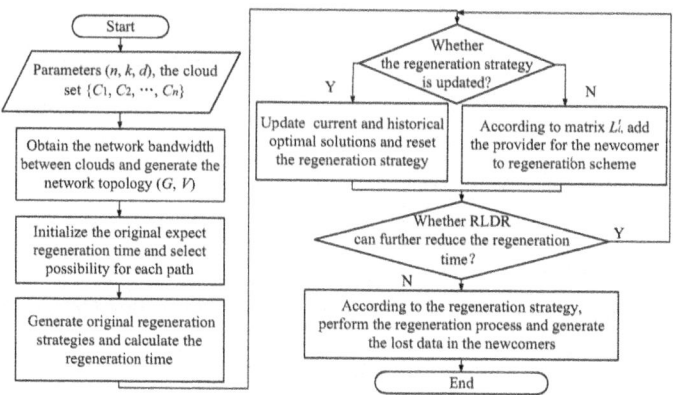

Fig. 5. Flow chart of RLDR.

4.2 Expect Regeneration Time Updating

To calculate the expect regeneration time for the topology, we adopt a Monte-Carlo based reinforcement learning algorithm. According to Eq. (14), RLDR can construct the regeneration tree by selecting the links with small expect regeneration time. Figure 4 illustrates the update process of expect regeneration time for a given network topology.

For agent a_i at time slot t, we record current optimal solution L_c^t to represent the optimal solution as detected by agent a_i, and historical optimal solution L_h^t to represent the optimal solution as detected among all the agents. Accordingly, we define

$$V_c^t = L_c^t - L_i^t, V_h^t = L_h^t - L_i^t, \quad (15)$$

where L_i^t is current expect regeneration time for agent a_i.

According to Eq. (15), RLDR needs to update expect regeneration time L_i^t. Inspired by Particle Swarm Optimization (PSO), we define velocity V_i^t for agent a_i as

$$V_i^{t+1} = w * V_i^t + \delta_c * r_c * V_c^{t+1} + \delta_h * r_h * V_h^{t+1}, \quad (16)$$

where w is the inertia weight, δ_c is the learning rate for current optimal solution, δ_h is that for historical optimal solution, and two random numbers $r_c, r_h \in (0,1]$.

According to Eq. (16), RLDR updates the matrix L_i^t by

$$L_i^{t+1} = L_i^t + V_i^{t+1}. \quad (17)$$

After updating the matrix L_i^t, the agent generates new regeneration tree. RLDR iteratively updates regeneration strategy according to network topology.

4.3 Reinforcement Learning Based Data Recovery

According to the expect regeneration time, we construct RLDR to perform the regeneration process. Figure 5 illustrates the flow chart of RLDR. Specifically, RLDR can perform the regeneration process on the following steps.

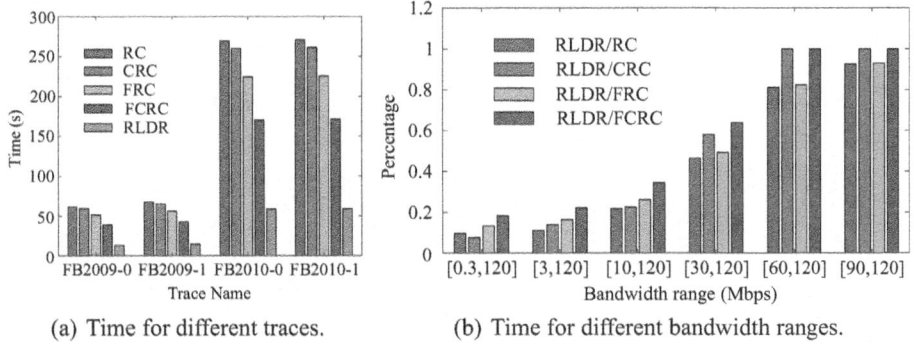

(a) Time for different traces.

(b) Time for different bandwidth ranges.

Fig. 6. Regeneration time under parameters $d = 5$, $k = 2$, and $r = 2$ by default.

First, RLDR generates the regeneration tree according to the network topology and coding parameters, and calculates the expect regeneration time for each regeneration strategies. Accordingly, RLDR can use the Monte-Carlo based reinforcement learning scheme to generate the regeneration tree.

Second, RLDR can determine the network traffic according to the available network bandwidth between clouds. According to Eqs. (6) and (9), we can calculate the data volume transferred from providers to newcomers and the network traffic between clouds for a regeneration tree.

Finally, the newcomers receive the data from providers and perform the regeneration process. By the Random Linear Code [8], RLDR can easily generate the lost data in the newcomers. Since RLDR can efficiently utilize the network bandwidth between clouds, the storage systems can quickly perform the regeneration process.

5 Experiments

5.1 Experimental Setup

We evaluate the performance of regeneration process with 7 storage nodes, so as to analyze how the network bandwidth between storage nodes affects the performance of regeneration process. Specifically, we conduct experiments with 6 bandwidth ranges: $U_1[0.3, 120]$ Mbps, $U_2[3, 120]$ Mbps $U_3[10, 120]$ Mbps, $U_4[30, 120]$ Mbps, $U_5[60, 120]$ Mbps, $U_6[90, 120]$ Mbps, and set $[10, 120]$ Mbps by default.

To simulate the file size in real-world data centers, we use Facebook SWIM traces [21] to evaluate the performance of recovery operations. We merge the small I/O requests to 256 MB to avoid small data volume transmission operations. We focus on two performance metrics (*i.e.*, regeneration time and throughput) [5]. The regeneration time is measured as the duration of regeneration process, while the throughput is measured as total recovered data volume divided by the duration of regeneration process.

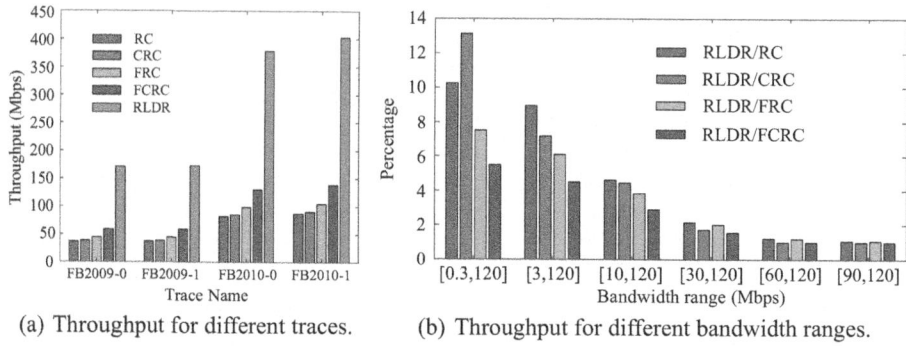

(a) Throughput for different traces.   (b) Throughput for different bandwidth ranges.

Fig. 7. Throughput of regeneration process under parameters $d = 5$, $k = 2$, and $r = 2$ by default.

5.2 Advantage of Our RLDR Scheme

First, we evaluate the performance of regeneration process under different traces. The results are shown in Figs. 6(a) and 6(b). In short, RLDR can reduce the regeneration time by up to 79% and improve the throughput by up to 400%, compared with the state-of-the-art alternatives. Meanwhile, RLDR can also quickly perform the regeneration process to enhance data reliability of storage systems.

Second, we describe the performance of regeneration process under different bandwidth ranges. The results are shown in Figs. 7(a) and 7(b). When the bandwidth range is large, RLDR can significantly improve the performance of regeneration process. Moreover, CRC needs to transfer given data volume between newcomers that is easily affected by the newcomers with low network bandwidth. When the bandwidth range is [0.3,120] Mbps, RLDR can reduce the regeneration time by up to 92%, and improve the throughput by up to 1210% compared with CRC.

6 Conclusion

Reducing data recovery overhead is the key issue in cloud-of-clouds storage systems. This paper proposed the RLDR scheme to reduce the regeneration time. By adopting Monte-Carlo based reinforcement learning, RLDR could efficiently generate regeneration tree to improve the performance of regeneration process. To accelerate the learning process, we directly calculated the optimal regeneration time according to available network bandwidth between clouds. We further implemented a prototype storage system to deploy various regeneration schemes. To verify the efficiency of RLDR, we conducted extensive experiments with real-world traces. The experimental results demonstrated that our scheme could efficiently reduce the regeneration time and improve the throughput of regeneration process, compared with the state-of-the-art alternatives.

References

1. Storer, M.W., Greenan, K.M., Miller, E.L., Voruganti, K.: POTSHARDS-A secure, recoverable, long-term archival storage system. ACM Trans. Storage (TOS) **5**(2), 5 (2009)
2. Li, L., Shen, J., Wu, B., Zhou, Y., Wang, X., Li, K.: Adaptive data placement in multi-cloud storage: A non-stationary combinatorial bandit approach. IEEE Trans. Parallel Distrib. Syst. **34**(11), 2843–2859 (2023)
3. Shen, J., Wu, B., Xiang, W., Zhao, Z., Zhang, K.: Adaptive network coding based disaster backup in cloud-of-clouds storage systems. In: 2023 15th International Conference on Communication Software and Networks (ICCSN), pp. 397–401 (2023)
4. Tang, H., Liu, F., Shen, G., Jin, Y., Guo, C.: UniDrive: synergize multiple consumer cloud storage services. In: Proceedings of the 16th Annual Middleware Conference (Middleware 15), pp. 137–148. ACM (2015)
5. Shen, J., Li, Y., Zhou, Y., Wang, X.: Mobile cloud-of-clouds storage made efficient: a network coding based approach. In: Proceedings of IEEE Symposium on Reliable Distributed Systems (SRDS 18), pp. 72–82 (2018)
6. Resch, J.K., Plank, J.S.: AONT-RS: Blending security and performance in dispersed storage systems. In: Proceedings of USENIX Conference on File and Storage Technologies (FAST 11), pp. 191–202 (2011)
7. Chen, H.C., Hu, Y., Lee, P.P., Tang, Y.: NCCloud: a network-coding-based storage system in a cloud-of-clouds. IEEE Trans. Comput. (TC) **63**(1), 31–44 (2014)
8. Dimakis, A.G., Godfrey, P.B., Wu, Y., Wainwright, M.J., Ramchandran, K.: Network coding for distributed storage systems. IEEE Trans. Inf. Theor. (TIT) **56**(9), 4539–4551 (2010)
9. Wang, Y., Wei, D., Yin, X., Wang, X.: Heterogeneity-aware data regeneration in distributed storage systems. In: Proceedings of IEEE Conference on Computer Communications (INFOCOM 14), pp. 1878–1886 (2014)
10. Speedtest. https://www.speedtest.net/
11. Shum, K.W., Hu, Y.: Cooperative regenerating codes. IEEE Trans. Inf. Theor. (TIT) **59**(11), 7229–7258 (2013)
12. Li, J., Yang, S., Wang, X., Li, B.: Tree-structured data regeneration in distributed storage systems with regenerating codes. In: Proceedings of IEEE Conference on Computer Communications (INFOCOM 10), pp. 1–9 (2010)
13. Pei, X., Wang, Y., Ma, X., Fu, Y., Xu, F.: Cooperative repair based on tree structure for multiple failures in distributed storage systems with regenerating codes. In: Proceedings of ACM International Conference on Computing Frontiers, pp. 1–8 (05 2015)
14. Shen, Z., Lee, P.P.C., Shu, J.: Efficient routing for cooperative data regeneration in heterogeneous storage networks. In: Proceedings of 2016 IEEE/ACM 24th International Symposium on Quality of Service (IWQoS 16), pp. 1–10 (2016)
15. Huang, C., et al.: Erasure coding in windows azure storage. In: Proceedings of the 2012 USENIX conference on Annual Technical Conference (ATC 12), pp. 15–26 (2012)
16. Shen, J., Gu, J., Zhou, Y., Wang, X.: Bandwidth-aware delayed repair in distributed storage systems. In: Proceedings of 2016 IEEE/ACM 24th International Symposium on Quality of Service (IWQoS 16), pp. 1–10 (2016)
17. Huang, J., Xie, C., Liang, X., Qin, X., Cao, Q.: PUSH: a pipelined reconstruction I/O for erasure-coded storage clusters. IEEE Trans. Parallel Distrib. Syst. (TPDS) **26**(2), 516–526 (2015)
18. Shah, N.B., Rashmi, K.V., Kumar, P.V.: A flexible class of regenerating codes for distributed storage. In: Proceedings of IEEE International Symposium on Information Theory (ISIT 10), pp. 1943–1947 (2010)

19. Shen, J., Gu, J., Zhou, Y., Xin, W.: Flexible regenerating codes for multiple node failures. In: Proceedings of CCF BigData Conference (CCF BigData 16), pp. 1–10 (2016)
20. Amazon EC2. https://aws.amazon.com/
21. Chen, Y., Alspaugh, S., Katz, R.: Interactive analytical processing in big data systems: a cross-industry study of MapReduce workloads. Proc. Very Large Data Base (VLDB) Endowment **5**(12), 1802–1813 (2012)

DAG: A Lightweight and Real-Time Edge Defense Model for IoT DDoS Attacks

Yanhua Liu[1,2], Cong Chen[1,2(✉)], Qiu Zhang[1,2], Fanhao Zeng[1,2], and Ximeng Liu[1,2]

[1] College of Computer and Data Science, Fuzhou University, Fuzhou 350116, China
[2] Fujian Provincial Key Laboratory of Networking Computing and Intelligent Information Processing, Fuzhou University, Fuzhou 350116, China
fzucckun@gmail.com

Abstract. Internet-of-Things (IoT) devices are increasingly used in people's lives and production in various industries. To detect and defend against Denial-of-Service (DDoS) attacks that occur on IoT networks, a lot of methods based on machine learning and deep learning have been proposed in recent years. However, these methods usually do not consider the limitation of computational resources of IoT devices. In this paper, we propose an edge model **D**DoS-**A**ttack-**G**uard (DAG) based on Bi-GRU and ShuffleNet for DDoS identification and classification with the target of lightweight and real-time. To demonstrate the performance of our models, we use the CICDDoS2019 dataset to test the identification and classification accuracy as well as the model inference time. In addition, we build a multi-layer coder-decoder structure that can extract the potential temporal contextual features of DDoS traffic, and introduce a reconstruction structure that can improve model training. Through ablation experiments and comparative experiments, our model has an average inference speed of 2.5 ms across different data sizes, which is 50% faster than the Sota method, while hitting 99.3% and 99.9% accuracy in identification and classification respectively.

Keywords: IoT security · DDoS Attacks · Edge Defense · Deep Learning · Reconstruction Structure

1 Introduction

The rapid development of emerging technologies such as sensors, smartphones, 5G communications, and virtual reality led to innovative applications such as connected industries, smart cities, smart energy, connected cars, smart agriculture, connected building complexes, connected healthcare, smart retail shops, and smart supply chains, the IoT has become an area with broad social impact and business potential, with latest research indicating that by 2025 there will be

---

This work was supported by the National Natural Science Foundation of China (Grant No.62072109, and No.U1804263), the Natural Science Foundation of Fujian Province (Grant No.2021J01625, No.2021J01616), Major Science and Technology project of Fujian Province (Grant No.2021HZ022007).

© The Author(s), under exclusive license to Springer Nature Singapore Pte Ltd. 2024
X. Wang et al. (Eds.): CCF ChinaNet 2023, CCIS 1988, pp. 61–73, 2024.
https://doi.org/10.1007/978-981-97-3890-8_5

5 trillion IoT devices [1]. IoT devices generate huge amounts of data in business communications creating a security risk for this vast network that cannot be ignored. According to Kaspersky Lab's collection, the number of malware samples for IoT devices has rapidly increased from 3,219 pieces in 2016 to 121,588 pieces in 2018, making it clear that there are a large number of vulnerabilities in IoT devices [2]. In general, these IoT devices are more vulnerable to attacks due to their limited computing power, storage resources, and network bandwidth compared to other devices in the network, such as smartphones, PC, edge servers, etc. [3].

The security requirements for confidentiality, integrity, availability, authentication, and access control make building a security defense for IoT devices unique and challenging [4], especially for developers, as developing a defense system that can withstand cyber attacks in a highly complex, resource-limited scenario requires thorough and well thought out and scenario-appropriate innovative mechanisms.

Great efforts and excellent contributions have been made in IoT security. As a precursor to deep learning, machine learning, which dynamically adapts to test environments after training on large data sets to obtain security models, can be used very effectively to analyze threats happening in the cybersecurity domain. However, traditional defense methods and machine learning algorithms consume large amounts of energy and are severely limited by complexity and compatibility in the IoT environment [4]. Moreover, many proposed security methods are highly centralized [5–10], which is not suitable for the IoT with its characteristics such as large-scale, peer-to-peer communication and susceptibility to single failures. However, existing methods of IoT security defense often face the following challenges:

1) Resource Constraints: most IoT devices have limited resources, including bandwidth, computation, and memory, which prevent efficient but complex solutions from being deployed on the device side [11].
2) Centralisation: the current IoT ecosystem is centered on cloud servers for device connectivity, communication, identification, and authentication. The systematically designed security mechanisms and defensive tools are centralized within the cloud server. Usually, one cloud server of an IoT service provider is often responsible for a specific commercial service, while tens of thousands of devices are connected. As a result, cloud servers act as a bottleneck and failure point in IoT security defense, and are prone to impact a large number of users in a network incident [12].
3) Time-sensitive: many organizations in high-priority sectors, such as government, energy, healthcare, banking, and research centers [13], necessitate strong network defense with an emphasis on confidentiality. In these sectors, the inability to efficiently detect and defend against zero-day attacks incurs substantial costs. However, existing tools and technologies prove inadequate in identifying novel attack types resulting from fluctuations in data volume, speed, diversity, and accuracy.

Hence, the IoT requires lightweight, distributed, and real-time security protection. The rapidly evolving deep learning techniques [14] in recent years are expected to overcome these challenges due to their powerful representational learning capabilities, unsupervised pre-training, and compression capabilities.

To address the above issues, we propose an identification-classification scheme based on network traffic and deep learning for IoT DDoS attacks. Our contributions can be briefly summarized as follows:

- We propose an end-to-end scheme DAG for the identification and classification of DDOS attacks. The DAG extracts the stream features from the raw traffic of the network and consists of a Bi-GRU and a ShuffleNet network, where the Bi-GRU module contains encoders, decoders, and a reconstruction structure.
- This reconstruction structure makes it easier to train features in a Bi-GRU network, improving model identification performance. And the classification results of ShuffleNet are fed back to the training of Bi-GRU as a probability distribution which can be treated as weights.
- Our DAG scheme lightens the network structure while maintaining 99.3% recognition accuracy and 99.9% classification accuracy, and improves the average inference speed of identification-classification task to 2.5ms per network traffic flow.

The rest of the paper is organized as follows. Section 2 describes the system modeling. The detailed architecture design is proposed in Sect. 3, and the experimental results are presented in Sect. 4. Finally, we conclude our work in Sect. 5.

2 System Modeling

This section explains the scheme's task definition, the sequence model, the ShuffleNet structure, and the self-encoder in turn.

2.1 Task Definition

The DDoS traffic defense problem in this paper is to identify and classify the DDoS attacks traffic into specific type using the sequences and images consisting of raw traffic information. In the first step, the Bi-GRU [15] network requires a sequence input, and the original network flows vary in length, so assume that a sequence of some length l_n is taken from each network flow as the input sequence. Next, we set the total number of samples to M, the attack types to C, so the t-th sample should be $x_t = [F_1^{(t)}, F_2^{(t)}, ..., F_{l_n}^{(t)}]$, where l_n is the length of x_t and F_i^t means the traffic value at time step i. Bi-GRU does the job of binary classification while ShuffleNet classifies the specific type of attack. Let attack label of x_t be $A_p(1 <= A_p <= C)$. Our target is to build a model $\psi(x_t)$ to predict a label $\hat{A}_p$ fitting A_p.

2.2 Sequence Model

RNN Model. The Recurrent Neural Networks (RNN) [16] are called 'loops' because they perform the same task for each element of the sequence, with the output depending on the previous computation. Another way of thinking about RNNs is that they have a 'memory' that holds the information they have calculated before.

Specially, given the input $x_t \in \mathbb{R}^m$, the output $o_t \in \mathbb{R}^n$ at time step t. The vanilla RNN model is calculated as below:

$$h_t = \tanh(W[h_t - 1, x_t] + b_1)$$
$$o_t = \text{softmax}(V \cdot h_t + b_2), \qquad (1)$$

where h_i and x_i contain the history information of time step i, (W, V, b_i) are updated during the training process.

Gated Recurrent Unit. GRU [15] aims to help solve the gradient vanishing problem in vanilla RNN by allowing the model to selectively retain or forget information from previous time steps. The advantages of GRU are shown below:

1) Better at capturing long-term dependencies than traditional RNNs, because the gating mechanism allows selective information retention and forgetting.

2) Requires less training time than other types of recurrent neural networks.

3) Has fewer parameters than an LSTM, making it faster to train and less prone to overfitting.

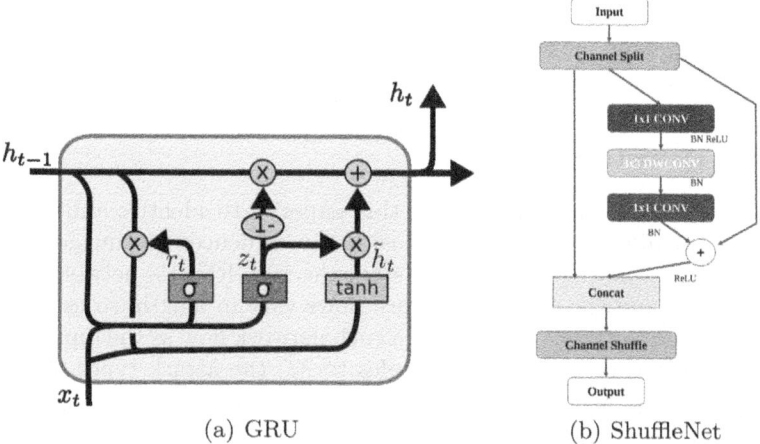

Fig. 1. 1(a) shows the architecture of gated recurrent unit, and 1(b) shows the structure of ShuffleNet

Figure 1(a) shows that the GRU cell contains an update gate and a reset gate. The output of the reset gate(the blue square) at time step t will be

$$r_t = \text{sigmoid}(W_r[x_t, h_{t-1}] + b_r), \tag{2}$$

where W_r and b_r are trainable parameters. Similarly, the output of the update gate(the purple square) is computed by

$$z_t = \text{sigmoid}(W_z[x_t, h_{t-1}] + b_z), \tag{3}$$

where W_z and b_z are weight matrix and bias learned from training. Hence, the hidden layer state can be updated as follows,

$$h_t = z_t \odot h_{t-1} + (1 - z_t) \odot \hat{h}_t, \tag{4}$$

where $\odot$ is the Hadamard product and the new memory information is shown below,

$$\hat{h}_t = \tanh(W_h[x_t, r_t \odot h_{t-1}] + b_h), \tag{5}$$

where W_h and b_h are also trainable parameters.

State h_t of the GRU can be updated from Eq.(2) to Eq.(5). The reset gate r_t determines how much the information of the previous state contributes to the current state and discards the irrelevant and useless information. The update gate controls the forgetting and updating of hidden states and helps the GRU cell to remember long-term information. Furthermore, the new state h_t is a linear interpolation of the previous state $h_t - 1$ and next state $\hat{h}_t$, which can avoid the gradient vanishing problem and model long-term memory. In the rest of the paper, we briefly describe the update process of the GRU hidden state using the following equation.

$$h_t = \text{GRU}(h_t - 1, x_t), \tag{6}$$

where h_t is the output of GRU at time step t.

2.3 ShuffleNet

As shown in Fig. 1(b), the ShuffleNet [17] is a lightweight cnn network that performs well in different fields. It is based on fewer parameters and a simpler networked structure. We apply it to the classification of DDoS attacks.

3 DDoS Attack Guard Scheme

The system view of the DAG is shown in Fig. 2. IoT DDoS attacks are usually initiated by network intruders who control a group of IoT devices called a botnet. The bots constantly generate service requests to cloud services, resulting in the exhaustion of network bandwidth and server resources. To solve this problem, we design DAG, a defense system deployed on edge-layer devices. The DAG accepts both upstream and downstream network flows, with the Bi-GRU module completing the binary classification of network flows and the ShuffleNet module completing the multi-classification of network flows. DDoS attack traffic will be discarded after being identified by the Bi-GRU. In addition, the classification results of ShuffleNet will be used as experience to improve the performance of Bi-GRU.

3.1 Bi-GRU Module

There are four parts to the Bi-GRU module shown in Fig. 3: Encoder, Decoder, Reconstruction, and Classification. We proposed this module by combining the advantages of supervised and unsupervised learning. In this section, we will describe the design principles and workflow in detail.

Traffic Flow Input. We take the network raw traffic as the model input, which can be defined as $x = [F_1, F_2, ..., F_{l_n}]$. F_i represents the value of each field in the network packet. The length l_n is a variable that means how much information the model utilizes in this one flow.

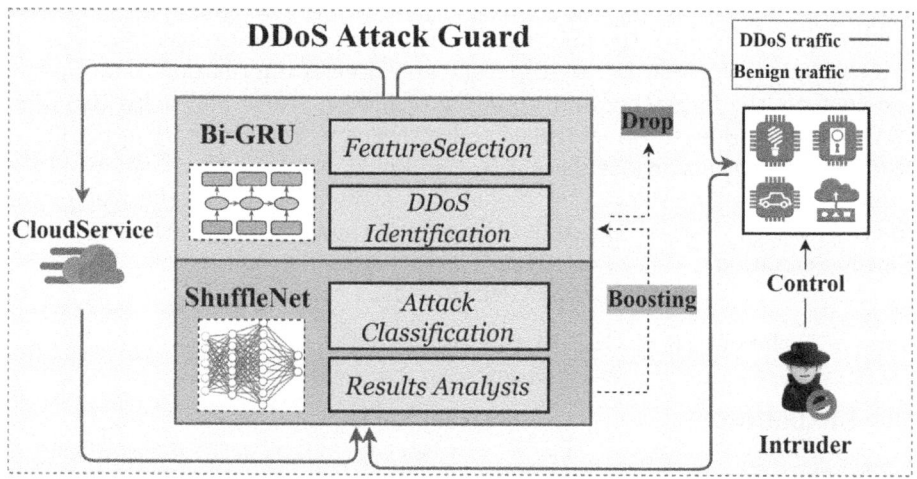

Fig. 2. The system overview of DAG

Encoder. The encoder takes the vectorized raw flows as inputs, then outputs the abstract and hidden features. It consists of a series of Bi-GRU units, so we obtain its output representation

$$\overrightarrow{h_t} = \overrightarrow{\text{GRU}}(\overrightarrow{h_{t-1}}, x_t), t \in [1, l_n] \tag{7}$$

$$\overleftarrow{h_t} = \overleftarrow{\text{GRU}}(\overleftarrow{h_{t-1}}, x_t), t \in [1, l_n], \tag{8}$$

where $\vec{h_t}$ and $\overleftarrow{h_t}$ are bi-directional hidden states of the encoder. Further, several Bi-GRU units form a layer in the neural network, and the encoder can consist of a multilayer Bi-GRU array. Hence, we denote the encoder output features as

$$o_e = [\vec{h_n}^{(k)}, ..., \overleftarrow{h_1}^{(k)}, ..., \vec{h_n}^{(1)}, ..., \overleftarrow{h_1}^{(1)}], \tag{9}$$

where k means the number of Bi-GRUs layers. Based on the Multi-layer design, the encoder can efficiently extract surface features and potential high-level features of traffic.

Decoder. In this paper, the decoder is a mirror of the encoder, they usually consist of Bi-GRU network layers of the same size. Similarly, we denote the decoder output as

$$\vec{d_t} = \overrightarrow{\text{GRU}}(\vec{d_{t-1}}, o_e), t \in [1, l_n] \tag{10}$$

$$\overleftarrow{d_t} = \overleftarrow{\text{GRU}}(\overleftarrow{d_{t-1}}, o_e), t \in [1, l_n] \tag{11}$$

$$o_d = [\vec{d_n}^{(k)}, ..., \overleftarrow{d_1}^{(k)}, ..., \vec{d_n}^{(1)}, ..., \overleftarrow{d_1}^{(1)}], \tag{12}$$

where $\vec{d_t}$ and $\overleftarrow{d_t}$ are bi-directional hidden states of decoder, k means the number of Bi-GRUs layer. We can obtain the size of the encoder output of high-level features trained by the decoder and reconstruct them back to the input traffic flow.

Reconstruction. The reconstruction layer fits the reconstructed output o_d of the decoder to the real data x, and we define the loss of this fitting process as

$$L_R = -\frac{1}{M}\sum_p^M \left|\hat{F}_p - F_p\right|, \tag{13}$$

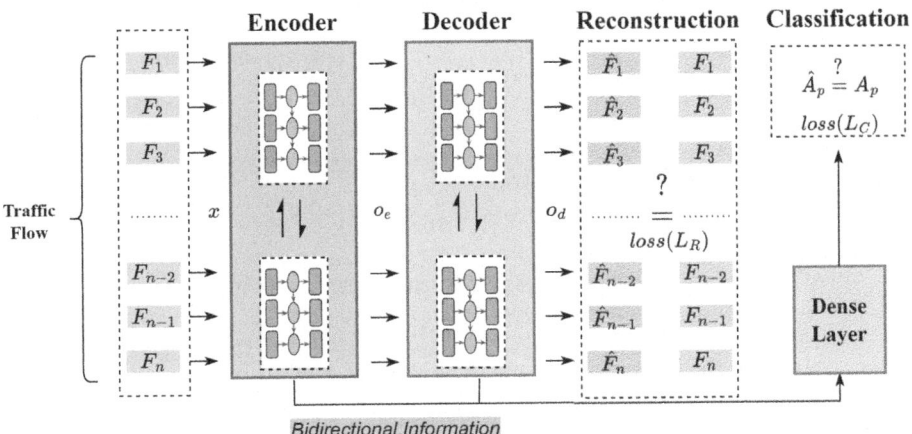

Fig. 3. The architecture of Bi-GRU

where $\left|\hat{F}_p - F_p\right|$ represents the difference between real data and reconstructed data. This loss can boost the encoder to learn effective features for classification.

Classification. The main component of the classification layer is the softmax classifier. Combining the encoder output o_e and the decoder output o_d, we get the input for classification

$$o = [o_e, o_d, |o_e - o_d|, o_e + o_d], \tag{14}$$

where $|o_e - o_d|$ is the difference between these two matrixs and $o_e + o_d$ is the element-wise plus of them. Since o is a combination of high dimensional data, it should be dimensionalized by the dense layer. To make the predicted label $\hat{A}_p$ close to the ground truth A_p, we define the loss of this part as

$$L_C = -\frac{1}{M}\sum_p^M -(\alpha \cdot y \cdot \log(\hat{y}) + \beta \cdot (1-y) \cdot \log(1-\hat{y})), \tag{15}$$

where $\hat{y}$ is the probability that the sample is DDoS attack traffic. (α, β) is the probability distribution computed after ShuffleNet classification. α indicates the probability that the sample traffic is a DDoS attack, while β is the opposite. The ShuffleNet classification results over a period of time are fed back to the Bi-GRU identification module as experience.

Model Training Procedure. Jointly with what is described in Sect. 3.1, the training procedure of Bi-GRU is shown in Algorithm 1.

Algorithm 1. Model Training Procedure

1: **procedure** BI-GRU($x_t, x_{t-1}, A_p, \hat{A}_p$)
2:     $(x_{t-1}, x_t) \leftarrow$ Network raw traffics at time step t and $t-1$
3:     **while** $(x_{t-1}, x_t) \neq 0$ **do**
4:         Compute $o_e =$ encoder(x_{t-1}, x_t), Contains bi-directional information $(\overrightarrow{h_t}, \overleftarrow{h_t})$
5:         Compute $o_d =$ decoder(o_e), Contains bi-directional information $(\overrightarrow{d_t}, \overleftarrow{d_t})$
6:         **if** Model in training **then**
7:             Compute $\hat{F}_p =$ Reconstruction(o_d)
8:             Fit $\hat{F}_p$ and F_p, define $L_R = -\frac{1}{M}\sum_p^M \left|\hat{F}_p - F_p\right|$
9:         **end if**
10:         Given classification input: $o = [o_e, o_d, |o_e - o_d|, o_e + o_d]$
11:         Compute $\hat{A}_p =$ Classification(o)
12:         Fit $\hat{A}_p$ and A_p, define $L_C = -\frac{1}{M}\sum_p^M -(\alpha \cdot y \cdot \log(\hat{y}) + \beta \cdot (1-y) \cdot \log(1-\hat{y}))$
13:         Update the network parameters based on L_R and L_C
14:     **end while**
15: **end procedure**

3.2 ShuffleNet Module

ShuffleNet has good image classification performance as a lightweight CNN network. We take part of the bytes of the raw traffic to generate formatted images and use them as the input of ShuffleNet. Usually, the loss of this classification network is defined as

$$L_S = -\frac{1}{M}\sum_{p}^{M}\sum_{c=1}^{C} f(A_p = c)\log\hat{y}_p(c), \qquad (16)$$

where $\hat{y}_p(c)$ is the propability of attack c for the sample x_p, and $f(A_p = c) = 1$ if $A_p = c$, else 0.

3.3 Loss

The loss of the DAG system is determined by L_C and L_R. We set a parameter to balance them, so the total loss of the DAG system is expressed as

$$L = L_C + \lambda L_R \qquad (17)$$

where λ is a variable that can be adjusted.

4 Evaluation

This section starts by introducing the experimental datasets, relevant metrics, and experimental settings. We then present the binary classification experimental results of the proposed DAG and compare them with other DDoS Detection methods. Finally, we compare DAG with other methods in terms of the time it takes.

As shown in Table 1, the CICDDoS2019 [18] dataset has a data size of more than 10 million containing all common attacks of DDoS. Considering the large size of the dataset and the limited computing power of our device, we sample 1/10 of the data of each category as training input and test model accuracy using ten-fold cross-validation.

To simulate the effect of deploying the DAG at the edge, the hardware resources of one PC are sufficient for our needs. The PC employed by our experiment is equipped with an Intel(R) Core(TM) i7-7700HQ CPU @ 2.80GHz and 16GB RAM. The version of Pytorch to deploy our models is 1.13.0. After amounts of experiments, we have verified that using the first 128 bytes of a network packet as input to the Bi-GRU module and the first 784 bytes as input to the ShuffleNet module will give the best results.

Table 1. CICDDoS2019 Dataset

Category	Type of Attack	Datasize
Benign	Benign	56863
Reflection-based attacks	NetBIOS	4093279
	MSSQL	4522492
	LDAP	2179930
	SSDP	2610611
	DNS	5071011
	SNMP	5159870
	NTP	1200642
	TFTP	20082580
Exploitation-based attacks	UDP	3134645
	UDP-lag	366461
	SYN	1582289

We evaluate the general performance of our model through four common machine learning evaluation metrics, *Accuracy*, *Precision*, *Recall*, and *F1 − Score*. Then, we verify the boosting effect of the reconstruction structure on model training through ablation experiments.

Table 2. Evaluation results of four models with CICDDoS2019

Type of Attack	ID3	RF	FlowGuard	DAG
Accuracy	-	-	-	**0.9981**
Precision	0.78	0.77	**0.9947**	0.9938
Recall	0.65	0.56	0.9931	**0.9942**
F1	0.69	0.62	0.9935	**0.995**

Table 2 illustrates the evaluation results of four models with CICDDoS2019. On the binary DDoS identification task, DAG achieves 99.83% in *Precision*, 99.62% in *Recall*, and 99.7% in *F1* compared to RF, ID3[18] and the latest FlowGuard[19]. Figure 4 shows the time it takes for the system to predict a raw traffic flow and complete the identification and classification task. When compared to the FlowGuard, the classification performance of our DAG is no worse than that of the FlowGuard, and it performs with an average runtime of 2.5 ms per flow. The average time required for FlowGuard to complete the task is 5.12ms. Therefore, our DAG has a significant improvement in runtime and better meets the real-time requirements in IoT defense.

To verify the effectiveness of the reconstruction module, we performed ablation experiments, and the model loss of the experimental process is shown in

Table 3. Experimental Results of ShuffleNet

Type of Attack	Benign	NetBIOS	MSSQL	LDAP	SSDP	DNS	SNMP	NTP	TFTP	UDP	UDP-lag	SYN
Accuracy	0.996	1	1	0.995	0.992	1	1	0.996	0.995	1	0.985	0.996
Precision	0.988	0.999	0.997	0.982	0.983	0.996	0.999	0.987	0.982	0.994	0.98	0.991
Recall	0.987	0.992	0.99	0.98	0.975	0.99	0.992	0.984	0.981	0.993	0.975	0.98
F1	0.983	0.994	0.992	0.978	0.98	0.991	0.995	0.982	0.978	0.991	0.978	0.987

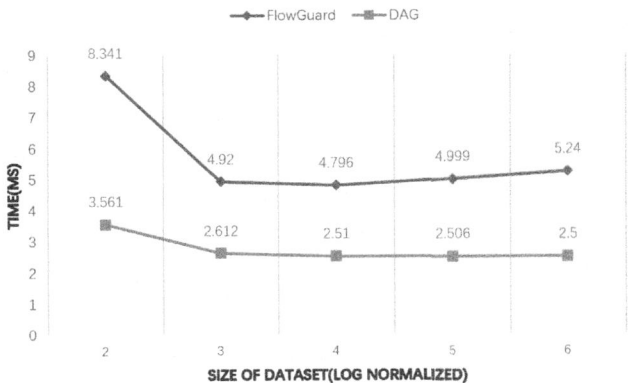

Fig. 4. The time comparison of DAG and FlowGuard[19]

Fig. 5. We reproduced the method used in the FlowGuard to obtain the training loss of LSTM+CNN in Fig. 5(a). Similarly, we logged the training loss of the DAG model and the DAG without reconstruction structure. They are presented in Fig. 5(b,c). After 100 epochs of training, the loss of the LSTM+CNN model converges to around 0.5 and still shows more pronounced fluctuations in subsequent training rounds. The loss of DAG without reconstruction structure also converges around 0.5, but its loss curve is less fluctuating. As shown in Fig. 5(c), the reconstruction structure helps the model to reduce the loss faster and more efficiently, and the final loss is around 0.13. Smoother loss curves imply higher stability of the model.

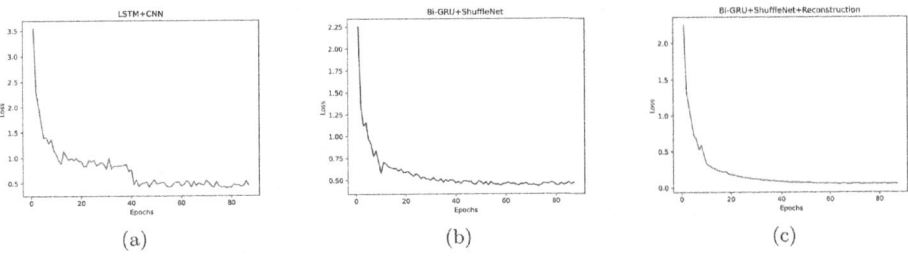

Fig. 5. Training loss of three models

Table 3 shows that our DAG also performs well on the DDoS attack multi-classification task. We conducted experiments on each of the twelve types of attacks included in the dataset, achieving an average score of 99.62% in *Accuracy*, 98.98% in *Precision*, 98.49% in *Recall* and 98.57% in $F1$. This performance is certainly excellent for ShuffleNet, which has fewer parameters and a simpler structure. At the same time, high-precision multiclassification results can boost the improvement of Bi-GRU results.

5 Conclusion

In this paper, we proposed an end-to-end DAG scheme for identifying and classifying DDOS attacks. The DAG extracts stream features from the original network traffic flow and consists of Bi-GRU and ShuffleNet networks, where the Bi-GRU module contains an encoder, a decoder, and a reconstruction structure. Specially, we proposed the reconstruction structure, which facilitates the training of features in a Bi-GRU network, improving model identification performance. Ablation experiments show the effectiveness of the reconstruction module. Furthermore, we fed the classification results of ShuffleNet into the training of Bi-GRU as a probability distribution with weights. Experimental results show that our DAG scheme simplifies the network structure while maintaining 99.96% Identification accuracy and 99.62% classification accuracy, and improves the average inference speed to 2.5 ms per network traffic flow. Compared with the Sota method, it is similar in *Arrcuracy*, while reducing about 50% of system runtime.

IoT defense mechanisms need to be constantly updated in the face of increasing IoT traffic data and potential zero-day attacks. In our future work, we aim to design a scenario-based model with generalization ability that will incorporate massive, heterogeneous, and multi-source IoT traffic data to defend against unknown attacks while satisfying the real-time and accuracy of traffic detection.

References

1. Kumari, P., Jain, A.K.: A comprehensive study of DDoS attacks over IoT network and their countermeasures. Comput. Secur. **127**, 103096 (2023)
2. Baranyi, P.: New trends in the world of iot threats. https://securelist.com/new-trends-in-the-world-of-iot-threats/87991/ (2019)
3. Bian, J.: Machine learning in real-time internet of things (iot) systems: a survey. IEEE Internet Things J. **9**(11), 8364–8386 (2022)
4. Waheed, N., He, X., Ikram, M., Usman, M., Hashmi, S.S., Usman, M.: Security and privacy in IoT using machine learning and blockchain: threats and countermeasures. ACM Comput. Surv. (CSUR) **53**(6), 1–37 (2020)
5. Narayanaswamy, B., Balaji, B., Gupta, R., Agarwal, Y.: Data driven investigation of faults in HVAC systems with model, cluster and compare (MCC). In: Proceedings of the 1st ACM Conference on Embedded Systems for Energy-Efficient Buildings, pp. 50–59 (2014)

6. Hayes, M.A., Capretz, M.A.: Contextual anomaly detection in big sensor data. In: 2014 IEEE International Congress on Big Data, pp. 64–71. IEEE, 2014
7. Lidong, F., Zhang, W., Tan, X., Zhu, H.: An algorithm for detection of traffic attribute exceptions based on cluster algorithm in industrial internet of things. IEEE Access **9**, 53370–53378 (2021)
8. Tamy, S., Belhadaoui, H., Rabbah, M.A., Rabbah, N., Rifi, M.: An evaluation of machine learning algorithms to detect attacks in SCADA network. In: 2019 7th Mediterranean Congress of Telecommunications (CMT), pp. 1–5. IEEE, 2019
9. Elnour, M., Meskin, N., Khan, K., Jain, R.: A dual-isolation-forests-based attack detection framework for industrial control systems. IEEE Access **8**, 36639–36651 (2020)
10. Alshammari, A., Zohdy, M.A.: Internet of things attacks detection and classification using tiered hidden markov model. In: Proceedings of the 2019 8th International Conference on Software and Computer Applications, pp. 550–554 (2019)
11. Ahmad, R., Alsmadi, I., Alhamdani, W., Tawalbeh, L.: A comprehensive deep learning benchmark for IoT IDS. Comput. Secur. **114**, 102588 (2022)
12. Amanullah, M.A.: Deep learning and big data technologies for IoT security. Comput. Commun. **151**, 495–517 (2020)
13. Habeeb, R.A.A., Nasaruddin, F., Gani, A., Hashem, I.A.T., Ahmed, E., Imran, M.: Real-time big data processing for anomaly detection: a survey. Int. J. Inf. Manage. **45**, 289–307 (2019)
14. Hao, X., Zhang, G., Ma, S.: Deep learning. Int. J. Semant. Comput. **10**(03), 417–439 (2016)
15. Cho, K.: Learning phrase representations using RNN encoder-decoder for statistical machine translation (2014). arXiv preprint arXiv:1406.1078
16. Sherstinsky, A.: Fundamentals of recurrent neural network (RNN) and long short-term memory (LSTM) network. Physica D **404**, 132306 (2020)
17. Ma, N., Zhang, X., Zheng, H.T., Sun, J.: Shufflenet v2: practical guidelines for efficient cnn architecture design. In: Proceedings of the European Conference on Computer Vision (ECCV), 116–131 (2018)
18. Sharafaldin, I., Lashkari, A.H., Hakak, S., Ghorbani, A.A.: Developing realistic distributed denial of service (DDoS) attack dataset and taxonomy. In: 2019 International Carnahan Conference on Security Technology (ICCST), 1–8 (2019)
19. Jia, Y., Zhong, F., Alrawais, A., Gong, B., Cheng, X.: Flowguard: an intelligent edge defense mechanism against IoT DDoS attacks. IEEE Internet Things J. **7**(10), 9552–9562 (2020)

A Data Publishing Method for Trajectory Privacy Classification Based on Differential Privacy

Qian He[✉], Bingjie Liao, Peng Liu, and Qinghe Dong

Guangxi Key Laboratory of Cryptography and Information Security, Guilin University of Electronic Technology, Guilin 541004, China
heqian@guet.edu.cn, 21032202019@mails.guet.edu.cn

Abstract. The traditional privacy protection method of trajectory data release protects the whole trajectory to the same degree, resulting in an unjustifiable allocation of the privacy budget and a diminishment in data accessibility. A technique is put forth to partition the privacy level of the trajectory based on the visitation frequency and duration, catering to the privacy preservation requirements of users with diverse data sensitivities. Density-based Spatial Clustering of Applications with Noise (DBSCAN) clustering is employed to categorize the trajectory points within high-density clusters into identical clusters, thus forming distinct trajectory clusters. The trajectory is segmented using the standard deviation to ensure uniform distribution of trajectory segments. This process streamlines the dataset, extracts distinctive behavioral patterns, and mitigates the spatiotemporal intricacies associated with trajectory data processing. A noisy trajectory segment prefix tree is assembled, and the privacy budget is allocated based on the weightage of the trajectory's privacy level and the tree's height. A Markov chain is introduced to constrain the magnitude of noise injected into the data. Experimental results substantiate the efficacy of the algorithm proposed in this paper in balancing data availability and privacy preservation.

Keywords: Trajectory privacy protection · Differential privacy · Privacy classification · Location Services · Noise prefix trees

1 Introduction

With the continuous development of Internet of Things (IoT) technology, various information data is experiencing an exponential growth rate. Social networks, diverse applications, and other entities avidly collect massive user data. Among these, trajectory data generated by mobile devices and vehicles has emerged as a significant source of information [1]. Trajectory data encompasses location, time, and speed, providing valuable insights for domains like traffic management, urban planning, and intelligent transportation systems [2, 3]. However, collecting, publishing, and sharing trajectory data may lead to privacy breaches for users [4]. This issue becomes particularly severe in social networks, e-commerce, and targeted advertising scenarios, where privacy concerns associated with trajectory data publication are heightened. Suppose trajectory data falls into the hands of malicious third-party organizations or hackers and is exploited for nefarious purposes. In that case, it can pose immeasurable risks and losses to users' well-being, financial security, and privacy.

Dwork et al. [5] introduced the concept of differential privacy in 2006 as a novel method for safeguarding privacy. Its fundamental idea revolves around the probability of obtaining identical results from neighboring datasets is remarkably close. Due to its resilience against background knowledge attacks and rigorous mathematical validation, differential privacy has garnered significant attention among privacy protection researchers in recent years and has found extensive application in trajectory disclosure.

This paper proposes a data publishing method for trajectory privacy classification based on differential privacy. This method protects trajectory privacy while satisfying the privacy requirements of different sensitivity levels of trajectories. The research content and main contributions of this paper are as follows:

1. Based on DBSCAN, we identify trajectory behavioral patterns and propose an algorithm for privacy level division based on access frequency and dwell time. By considering the attributes of stay points and hotspots, determine the sensitivity level of trajectory points and classify trajectories into three privacy levels to meet diverse user needs.
2. We construct a noise trajectory segment prefix tree and allocate a privacy budget reasonably based on the tree height and privacy level weights. Laplace noise is added to the prefix tree nodes, and a Markov chain is introduced to limit the magnitude of the noise, ensuring maximum trajectory usability while preserving trajectory privacy.
3. Accurate trajectory datasets are used for experimentation to evaluate the effectiveness of the proposed method in this paper. The results demonstrate that the method presented in this paper effectively balances the usability and privacy of trajectory publishing.

The remainder of the paper is organized as follows. Section 2 discusses the related work. Section 3 presents the methods and implementation process of trajectory data publication based on differential privacy. Section 4 conducted experimental analysis, while Sect. 5 presented the findings.

2 Related Work

Regarding the issue of privacy protection in trajectory publishing, numerous effective methods have been proposed by scholars both domestically and internationally.

In [6], the trajectory space is partitioned into multiple planes using timestamps, then reconstruction of new trajectories through clustering and generalization techniques. Mechanism-based differential privacy is proposed to address privacy leakage issues in vehicular location data sharing. However, this approach sets a fixed value for the length of trajectories. [7] employs the exponential and Laplace mechanisms of differential privacy to introduce dual random perturbations to the data while utilizing the k-means clustering algorithm for sampling data processing. In [8], the Dijkstra method is combined with the minimum description length principle to reduce the spatiotemporal complexity of trajectory data. Furthermore, Laplace noise is added to the nodes of trajectory segment prefix trees to ensure trajectory privacy. This method introduces the concept of trajectory segment prefix trees for the first time. [9] employs the Hilbert curve to partition trajectory points and significantly enhances query efficiency by storing aggregated trajectory points in prefix trees. [10] proposes a differential privacy-based approach for protecting activity patterns and spatiotemporal data release. It regulates the dynamic and static information through privacy protection budgets and thresholds, subsequently dividing the spatiotemporal grid based on the regulated information.

However, traditional differential privacy protection methods have not considered users' privacy preferences, thus failing to meet their diverse needs. Therefore, in [11], an algorithm for dividing sensitive road privacy levels based on road network topology is proposed. It sets an initial sensitive location and determines the sensitivity of user locations based on their distance from the initial sensitive location. This is achieved by allocating privacy budgets to sensitive road segments and adding Laplace noise to protect location data privacy. In [12], users are assumed to have different privacy requirements and a trajectory data generalization method based on privacy budget weighting is proposed. Sampling and exponential mechanisms are introduced in generating representative elements for location clusters to satisfy users' personalized privacy preferences. However, this method randomly divides trajectories into three privacy levels to achieve personalized privacy protection mechanisms. In [13], to address the issue of location service privacy leakage in vehicular networks, the utility model is constructed using multi-attribute theory to allocate appropriate privacy budgets to users based on distance proportions as a measure. Users' privacy preference requirements are integrated into this model. However, areas where users frequently and for extended periods stay, contain more privacy information and are more likely to expose users' behavioral patterns. Existing personalized privacy protection methods have not considered this, resulting in trajectory privacy leakage. This paper proposes a trajectory privacy level classification method based on the number of visits and residence time of the trajectory to meet the privacy protection requirements of users for data with different sensitivity.

3 Trajectory Data Publishing Method

3.1 Methodology and Procedure

The trajectory publication process diagram, as depicted in Fig. 1, encompasses three main modules: the data acquisition module, the data processing module, and the privacy protection module. The data acquisition module primarily retrieves user locations through GPS positioning and stores them in a database. The data processing module comprises preprocessing, trajectory segmentation, and trajectory privacy level classification. Initially, the data is cleansed and segmented into trajectory segments, followed by the classification privacy levels. The privacy protection module first stores the data segments into the prefix tree nodes and assigns noise to the prefix tree nodes to ensure the balance between data availability and privacy.

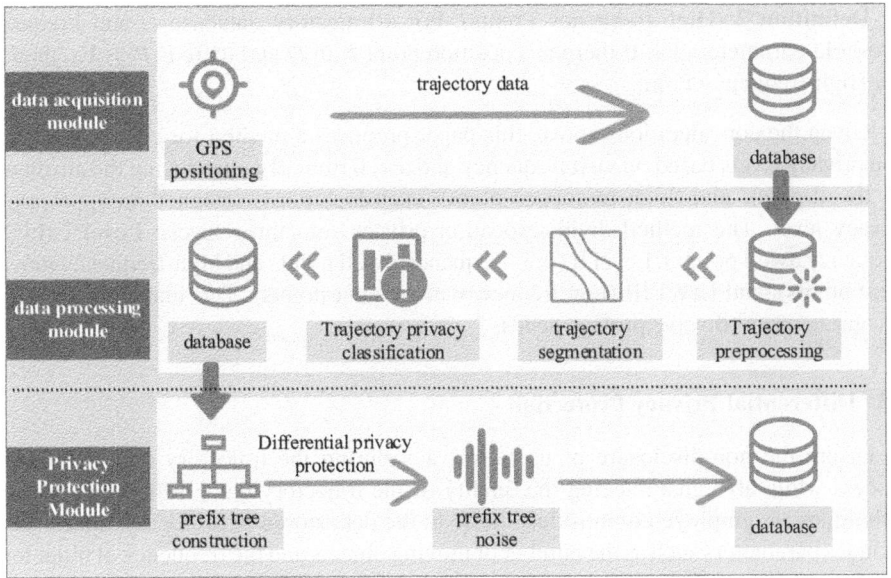

Fig. 1. Trajectory publication privacy protection flow chart.

3.2 Trajectory Privacy Classification

Trajectories are not merely a collection of randomly sampled spatiotemporal points; they often possess specific semantics, particularly those trajectory points that are frequently accessed or have longer durations of stay. These trajectory points pertain to an individual's activity path, thus making them more susceptible to exposing personal information and becoming potential attack targets. Dwell points refer to location points within a trajectory with a continuous and extended period of stay. Their definition is as follows:

Definition 1 (Dwell Point). [14] For a trajectory $T_i = \{l_{i,j} | j = 1, 2, ..., |T_i|\}$, time threshold parameter $time_\theta$, and distance threshold parameter $dist_\theta$, if conditions $1 \leq m \leq n \leq |T_i|$ $n - m > time_\theta$ are satisfied at the same time:

$$l_{i,m} = l_{i,m+1} = ... = l_{i,n} = l_{stay}$$

$$dist_{[l_{i,m},l_{i,m+1}]} \leq dist_\theta, dist_{[l_{i,m+1},l_{i,m+2}]} \leq dist_\theta, ..., dist_{[l_{i,n-1},l_{i,n}]} \leq dist_\theta$$

l_{stay} is a dwell point.

Furthermore, locations frequently visited by users belong to high-frequency areas. Although the high-frequency points do not directly expose users' sensitive information, attackers can analyze users' access frequency to obtain the high-frequency areas of user activities and subsequently disclose more covert information.

Definition 2 (High-frequency Points). For a trajectory database D and Frequent threshold parameters V_θ, If there is a position point P in D and there is $P \geq V_\theta$, then P is a High-frequency Point.

Given the considerations above, this paper proposes a method for classifying location privacy levels based on visit frequency and dwell time. By introducing the attributes of 'dwell point' and 'high-frequency point', each location is assigned an appropriate privacy level. The method divides location privacy into three levels: Level I (high-frequency dwell points), Level II (low-frequency dwell points and high-frequency movement points), and Level III (low-frequency movement points). This classification aims to cater to users' diverse privacy needs.

3.3 Differential Privacy Protection

To ensure the non-disclosure of users' privacy during the trajectory data publishing process while also guaranteeing the quality of the trajectory data, differential privacy techniques are employed to introduce noise to the data stored in prefix tree nodes. This includes parameters such as the number of moving objects and the frequency of trajectory segments. Markov processes and privacy level weights are also utilized to constrain the added noise.

In this paper, Laplace noise is added to the count results of each node in the prefix tree, and the privacy budget consumed is shared among nodes at the same level. The total privacy budget is denoted as ε, and the allocation of the privacy budget for each level of nodes is as follows:

$$\varepsilon_i = \frac{(h-i+1)i\varepsilon}{h(h+1)} \cdot \frac{1}{\sum_{i=1}^{h} \frac{(h-i+1)i}{h(h+1)}}$$

where h represents the height of the prefix tree, the total privacy budget consumed in the prefix tree is denoted by $\varepsilon = \sum_{i=1}^{h} \varepsilon_i$. By employing this allocation method, it is possible to distribute the privacy budget in a manner that maximizes its utilization, thereby avoiding situations where lower-layer nodes have insufficient privacy budget while higher-layer nodes have an excessive privacy budget, thus maintaining a reasonable balance. Additionally, data availability can be ensured by incorporating Markov processes and privacy level weights to constrain noise.

Procedure 1 Trajectory Privacy-Preserving Algorithm(STNV)
Input: total privacy budget ϵ, Trajectory dataset D_c, Markov probability P_{ij}, Privacy level weights W_{ij}
Output: Privacy Preserved Trajectory Dataset D_ϵ
1. Create a prefix tree
2. $i=1, \epsilon = \sum_{i=1}^{h} \epsilon_i$
3. calculate $sum(h) = \sum_{i=1}^{h} \frac{(h-i+1)i}{h(h+1)}$
4. $\epsilon_i = \frac{(h-i+1)i\epsilon}{h(h+1)} \cdot \frac{1}{sum(h)}$
5. for $i <= h$ do
6. $\quad j=1$
7. $\quad$ When every node of the i-th layer of the prefix tree exists
8. $\quad\quad tr_{ij} = tr_{ij} + P_{ij} \times W_{ij} \times laplace(\frac{1}{\epsilon_i})$
9. $\quad\quad j = j+1$
10. $\quad$ end
11. $\quad i = i+1$
12. end
13. return D_ϵ

First, construct a prefix tree and define the data structure of the prefix tree node for trajectory segments as $< tr, P(tr), O(tr), N(tr) >$. Here, tr represents the trajectory segment, $P(tr)$ denotes the privacy level of that trajectory segment. It stipulated in this paper that the trajectory point with the highest privacy level in each node's trajectory segment will be considered the prefix tree's trajectory segment privacy level. $O(tr)$ represents the number of moving objects on the trajectory segment while $N(tr)$ indicates the frequency of occurrence of that trajectory segment.

Store trajectory segment information in each node of the prefix tree and allocate a privacy budget based on the height of the prefix tree. Allocate the same privacy budget to each layer of nodes, with varying privacy budgets as the height of the prefix tree changes. Next, traverse the tree using a hierarchical approach to visit each node and collect node information. Inject Laplace noise into the data stored in each node. Since each layer of the prefix tree consists of several non-overlapping trajectories from the original trajectory, it can be inferred from the parallel composition property of differential privacy that nodes on the same level share the privacy budget consumed. The privacy budget consumption for each layer node is denoted as $\varepsilon_i = \frac{(h-i+1)i\varepsilon}{h(h+1)} \cdot \frac{1}{sum(h)}$, and the total privacy budget consumed is denoted as $\varepsilon = \sum_{i=1}^{h} \varepsilon_i$. The overall algorithm consumes a privacy budget equal to ε. Therefore, the algorithm involved in constructing the prefix tree for noisy trajectory segments satisfies the ε - differential privacy.

4 Experimental Analysis

4.1 Experimental Datasets and Environments

This paper uses the GPS trajectory dataset from T-Drive data [15], released by Microsoft Research, to evaluate algorithm performance. The dataset consists of the trajectories of 10,357 taxis in Beijing, China, between February 2nd and 8th, 2008. Each trajectory record includes the taxi ID, timestamp, and current location (longitude and latitude). The original data format is shown in Table 1:

Table 1. T-Drive dataset format.

ID	Time	Longitude	Latitude
1	2008-02-02 15:36:08	116.51172	39.92123
2	2008-02-04 19:31:51	116.45565	39.90723
3	2008-02-06 10:47:54	116.58383	39.91903

The experimental environment of this study consisted of Windows 10, Intel® Core™ i7-8700K CPU @ 3.70 GHz, and 32 GB of memory. The implementation was done using Python 3.8.

4.2 Algorithmic Analysis

The paper analyzes the feasibility and effectiveness of algorithms from two major perspectives: data availability and privacy. It introduces Definition 2 as evaluation criteria.

Definition 3 (Average Relative Error) [16]: The average relative error is calculated by assessing the data usability between the processed and original datasets through counting queries.

$$\text{Average} - \text{relative} - \text{Error} = \frac{|Q(D\prime) - Q(D)|}{\max\{Q(D), thr\}}$$

One of the purposes of a variable thr is to establish a threshold to prevent excessive selectivity in Q the query, thereby avoiding division by zero.

To validate the privacy and data availability of the algorithm proposed in this paper, comparative experiments were conducted with the NTPT algorithm proposed by Zhao et al. [8] and the TDPP algorithm proposed by Wu et al. [9].
(1) Availability analysis
The paper analyzes data availability, examining the impact of privacy budgets on average relative error. Figure 2 analyzes the effect of privacy budgets on relative error under varying query lengths. In this study, the term "query length" refers to the specified length of trajectories, and two sets of experiments were conducted based on different query lengths. Since this experiment's stored trajectory segment length is 9, the tree's

height is also set to 9. Therefore, the query lengths are set as 4 and 8, respectively. Each query object is obtained from a dataset of motion object trajectories, and the experiment is repeated 10 times, with the average of the 10 results taken as the outcome.

As shown in Fig. 2, it can be observed that regardless of the query length, the average relative error decreases as the privacy budget increases. This is because the noise added to the prefix tree nodes decreases as the privacy budget increases. In other words, the gap between the noisy data and the original information also decreases, resulting in increased data accuracy. Additionally, with an increase in the query length, there is a decreasing trend in the average relative error. This is due to the rise in the height of the prefix tree caused by the longer queries. The constraints imposed by the Markov chain and privacy level weights reduce the noise on the lower-layer nodes, lowering the average relative error. Figure 2 shows that the STNV algorithm reduces the average relative error by 35% compared to the NTPT algorithm and by 15.7% compared to the TDPP algorithm. The algorithm consistently exhibits lower errors than the NTPT and TDPP algorithms, indicating higher utility.

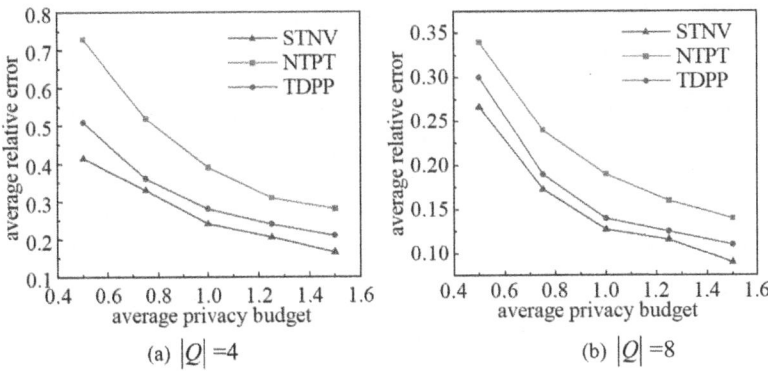

Fig. 2. Average relative error for different privacy budgets.

(2) Privacy analysis

This paper analyzes the privacy evaluation by examining the average relative error under different privacy budgets. Figure 3 shows that the average relative error decreases under the same privacy budget as the tree height increases. Additionally, increasing the privacy budget reduces the average relative errors when the tree height remains constant. The analysis demonstrates that controlling the size of the privacy budget ensures the availability of trajectory publishing and safeguards the privacy of trajectory data.

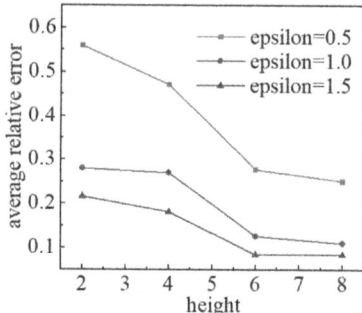

Fig. 3. Average relative error for different privacy budgets.

5 Conclusions

Based on the sensitivity of different trajectories, this paper proposes a trajectory privacy protection method for trajectory publishing. By designing an algorithm that divides trajectories into different privacy levels based on access frequency and dwell time, the trajectories are categorized according to their respective privacy levels. Additionally, a Markov chain and privacy level weights are employed to limit the added noise, thereby achieving protection of the trajectories. Furthermore, a reasonable privacy budget is allocated to each layer node of the prefix tree based on its height, ensuring a moderate balance of privacy budgets within the prefix tree. Experimental results demonstrate that the data publishing method presented in this paper can ensure both trajectory privacy protection and improved data availability.

Acknowledgment. This work is partly supported by the National Natural Science Foundation of China (62162018), The Natural Science Foundation of Guangxi (No. 2019GXNSFGA245004) Funding.

References

1. Jiang, B., Li, J., Yue, G., et al.: Differential privacy for industrial Internet of Things: opportunities, applications, and challenges. IEEE Internet Things J. **8**(13), 10430–10451 (2021)
2. Wang, S., Bao, Z., Culpepper, J.S., et al.: A survey on trajectory data management, analytics, and learning. ACM Comput. Surv. (CSUR) **54**(2), 1–36 (2021)
3. Feng, D., Zhang, M., Ye, Y.: Research on location trajectory publishing technology based on differential privacy model. J. Electron. Inf. Technol. **42**(1), 74–88 (2020). (in Chinese)
4. Ma, Z., Zhang, T., Liu, X., et al.: Real-time privacy-preserving data release over vehicle trajectory. IEEE Trans. Veh. Technol. **68**(8), 8091–8102 (2019)
5. Dwork, C., Kenthapadi, K., McSherry, F., Mironov, I., Naor, M.: Our data, ourselves: privacy via distributed noise generation. In: Vaudenay, S. (eds.) Advances in Cryptology - EUROCRYPT 2006. EUROCRYPT 2006. LNCS, vol. 4004, pp. 486–503. Springer, Heidelberg (2006). https://doi.org/10.1007/11761679_29

6. Xu, J., Liu, L., Zhang, R., et al.: IFTS: a location privacy protection method based on initial and final trajectory segments. IEEE Access **9**, 18112–18122 (2021)
7. Chen, S., Fu, A., Su, M., et al.: Trajectory privacy protection scheme based on differential privacy. J. Commun. **42**(9), 54–64 (2021). (in Chinese)
8. Zhao, X., Pi, D., Chen, J.: Novel trajectory privacy-preserving method based on prefix tree using differential privacy. Knowl.-Based Syst. **198**, 105940 (2020)
9. Wu, W., Zhao, Y., Wang, Q., et al.: A secure storage and publishing method for trajectory data satisfying differential privacy. Comput. Res. Dev. **58**(11), 2430–2443 (2021). (in Chinese)
10. Zeng, Z., Wang, C., Ma, F.: Differentially private activity pattern and spatial-temporal trajectory publication. Acta Electron. Sin. **51**(3), 552–563 (2023). (in Chinese)
11. Li, H., Ren, X., Wang, J., et al.: Continuous location, privacy protection mechanism, based on differential privacy. J. Commun. **42**(8), 164–175 (2021). (in Chinese)
12. Tian, F., Wu, Z., Lu, L., et al.: A sample based personalized differential privacy mechanism for trajectory data publication. Chin. J. Comput. **44**(04), 709–723 (2021). (in Chinese)
13. Xu, C., Din, Y., Luo, L., et al.: Personalized location privacy protection for location-based services in vehicular networks. J. Softw. **33**(02), 699–716 (2022). (in Chinese)
14. Huo, Z., Meng, X., Hu, H., et al.: You Can Walk Alone: trajectory privacy-preserving through significant stays protection. In: Lee, Sg., Peng, Z., Zhou, X., Moon, Y.S., Unland, R., Yoo, J. (eds.) Database Systems for Advanced Applications: 17th International Conference, DASFAA 2012, Busan, South Korea, 15–19 April 2012, Proceedings, Part I 17, vol. 7238, pp. 351–366. Springer, Cham (2012). https://doi.org/10.1007/978-3-642-29038-1_26
15. Yuan, J., Zheng, Y., Zhang, C., et al.: T-drive: driving directions based on taxi trajectories. In: Proceedings of the 18th SIGSPATIAL International Conference on Advances in Geographic Information Systems, California USA, pp. 99–108 (2010)
16. Xiao, X., Bender, G., Hay, M., et al.: iReduct: differential privacy with reduced relative errors. In: Proceedings of the 2011 ACM SIGMOD International Conference on Management of data, Athens Greece, pp. 229–240 (2011)

Towards Anomaly Traffic Detection with Causal Interpretability Methods

Zengri Zeng[1,3,4], Baokang Zhao[2(✉)], Xuhui Liu[3], and Xiaoheng Deng[4]

[1] Hunan University of Humanities, Science and Technology, Loudi, Hunan, China
[2] National University of Defense Technology, Changsha, Hunan, China
bkzhao@nudt.edu.cn
[3] Hunan Valin Lianyuan Iron, Loudi, Hunan, China
[4] Central South University, Changsha, Hunan, China

Abstract. The large non-independent and identically distributed (N-IID) samples result in a lack of stability and causal interpretability in the detection results of existing detection methods. To solve these problems, we propose an abnormal traffic detection method based on causal interpretability. This method first removes the false associations between features through Fourier feature transformation. Subsequently, a structural causal model (SCM) is constructed and pruned based on causal effects, and counterfactual diagnosis, thereby restoring the causal relationship between abnormal labels and traffic features. Verification on the CICIDS2019 and ToN_IoT datasets shows that this method effectively removes noise features, redundant information and false associations to effectively restore the causal relationships between network attacks and abnormal traffic features, ensuring good detection precision, guaranteeing detection stability when traffic is polluted and causal interpretability for network anomalies.

Keywords: network security · anomaly traffic detection · causal interpretability · SCM pruning

1 Introduction

Compared with traditional devices such as computers, the biggest difference in existing internet devices is that they are small in size and can connect to multiple devices; thus, there are more ways to attack them [1]. For example, according to the "2023 Global Automotive Industry Cybersecurity Report" released by the upstream research firm, the global automotive industry has suffered losses of more than 500 billion US dollars due to cyberattacks in the past five years, posing a very serious threat to human travel safety. Although previous anomaly detection methods seem to alleviate the harm caused by network anomalies to some extent [2], they have two unresolved problems. First, the differences in network and device types, configurations, application environments, etc., cause the training data and detection data to be N-IID, which greatly damages the stability of the detection model. Furthermore, although the development of technologies such as machine learning has improved the ability of models to detect network anomaly traffic, discriminating based on correlation methods confuses causal correlation and false correlation. This approach not only cannot provide causal traceability for traffic anomalies but also may lead to false conclusions and destructive consequences due to strong false correlations. For example, in the field of network abnormal traffic detection with prior knowledge, due to the interference of factors such as noise, exogenous latent variables and confounding factors, algorithm-level methods can confuse network attack labels and traffic features, leading to false associations, reducing the robustness of detection methods, and decreasing causal interpretability of detection results [3].

In recent years, with the emergence and development of causal reasoning technology, an effective method has been developed to reveal the causal relationships among machine learning methods and other methods. Cui Peng [4] suggested that using causal reasoning to guide machine learning can eliminate the instability impact of N-IID samples by eliminating false correlation. Second, causal interpretable artificial intelligence has become increasingly important for using deep learning and other models to enhance trust management [4, 5]. Therefore, to address the instability and causal interpretability problems of anomaly detection, we propose toward anomaly traffic detection with causal interpretability methods (ADCIM). This method first eliminates false associations through Fourier feature transformation and uses a Bayesian network to construct the SCM of network attacks. Next, based on the connections and differences between correlations and causalities, the SCM is pruned from the perspective of network attack labels. Finally, on the pruned SCM, the causal correspondence between each attack label and feature is clarified through counterfactual diagnosis methods, allowing the detection results to be causally interpretable. The main contributions of this paper are as follows:

(1) By exploring the average difference between intervention and nonintervention through causal intervention, the causal impact of features on network attack is estimated, noise features are identified and removed.
(2) Based on the pruned SCM, counterfactual diagnosis is used to restore a single network attack that has a true causal relationship with the abnormal feature.

2 Literature Review

As the system boundary of the existing internet devices continues to expand, it no longer is only a traditional hardware product but also involves organically integrates clouds and sensing nodes/edge nodes. Therefore, the existing internet devices faces security problems and challenges that are significantly different from those of traditional network environments [6]. To address these challenges, Dwivedi [7] proposed an intrusion detection system based on multiple artificial intelligence algorithms. The system uses features such as messages, timestamps, and packets to identify potential network attacks and considers time series features and attack frequency. Jiang [8] proposed a new machine audio anomaly detection and localization method based on a generative adversarial network (GAN), which is a completely unsupervised method that effectively addresses unknown network anomalies in the existing internet devices. In the era of the Internet of Things, traditional anomaly detection techniques based on distance and dimension cannot be used to effectively detect point anomalies with periodicity in stream data. Wang [9] proposed a new time series anomaly detection method based on deep learning (DeepAnT);

this method is applicable to both stream and nonstream data. Van [10] used a Kalman filter detector combined with deep learning methods to detect and identify abnormal sensor behaviors in CAVs. The method uses a CNN to extract features from sensor data and inputs them into the Kalman filter to estimate the sensor state. These studies all reflect the enormous potential and prospects of deep learning technology in anomaly detection.

3 Interpretable Anomaly Detection with SCM Construction and Pruning

3.1 Introduction to the Holistic Approach

In this chapter, a method for ADCIM is developed that can restore causal correspondence between network attacks and abnormal traffic features and thus achieve fast and accurate detection of network attack types. As shown in Fig. 1, the method consists of four main stages.

The first stage is by random Fourier feature mapping, the traffic features are mapped to a high-dimensional space to eliminate the linear correlation between the new features and ensure strict independence between the original features. Finally, based on the Bayesian network model, a preliminary network attack SCM is constructed according to the preprocessed data, as shown in Fig. 2a.

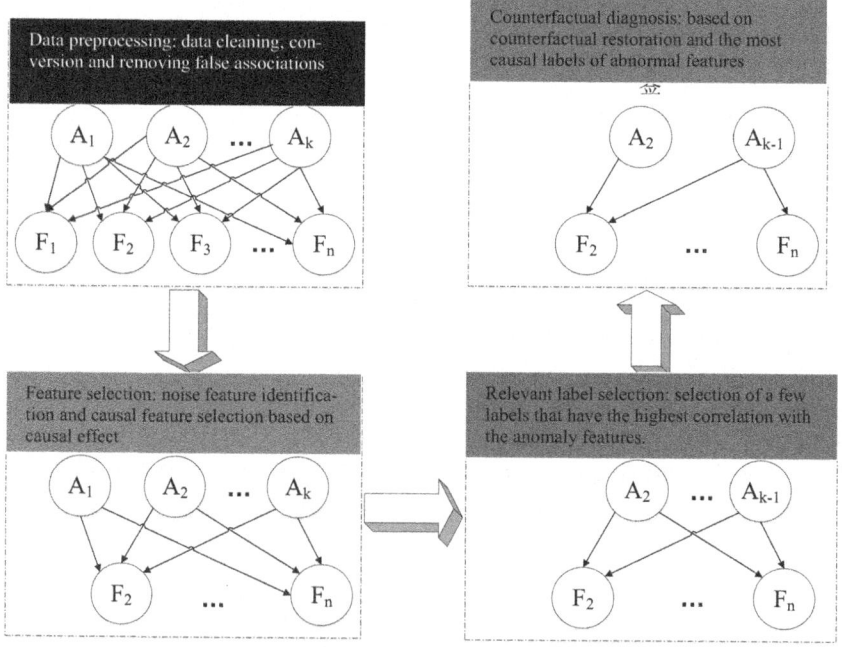

Fig. 1. ADCIM construction and pruning process

The second stage is feature selection based on causal effects (FSBCE), which reduces the number of features needed by machine learning and counterfactual diagnosis algorithms and improves the detection rate. First, by stacking weights on the features, the causal features and noise features are kept independent. These noise features are then deleted to prune the SCM. This method simplifies the correspondence between network attacks and features, reduces the time complexity of the detection method and improves the detection efficiency.

In the third stage, several attack labels that are most relevant to abnormal features are selected through machine learning algorithms to prune the SCM. If there is no correlation between the result labels and features, there must be no causal relationship; however, if there is a causal relationship between them, there must be a correlation. In this stage, by combining the three principles of causal inference, several result labels that are most related to the abnormal features are detected based on correlation inference, and the SCM is pruned accordingly, thereby reducing the algorithmic complexity of the fourth stage.

In the fourth stage, the expected values of abnormal features in the counterfactual dual network are calculated for K types of attacks in the third stage based on counterfactual diagnosis. A larger expected value of a certain type of attack indicates that the attack is more likely to cause abnormal features, and vice versa. At this point, the pruning and construction of the SCM between network attacks and features are completed, and the causal correspondence between various network attacks and abnormal features is restored.

Under the joint action of these four stages, when the network attacks increase, ADCIM can effectively restore the causal correspondence between network attacks and abnormal features and ensure a high detection rate.

3.2 Remove the False Associations Between Features

Definition 1 (Noise feature). Suppose that $F = \{F_c, F_n\}$ is the network traffic feature set, where F_c is the causal feature set, F_n represents the noise feature set $F_n = F \setminus F_c$, and $A \in \{0, k\}$ represents the network attack, totaling k types. Since there is no causal relationship between the noise feature F_n and network attack A, the conditional probability $Pr(A|F)$ satisfies the following condition [11]:

$$Pr(A|F) = Pr(A|F_c, F_n) = Pr(A|F_c) \qquad (1)$$

The false correlation between the noise feature F_n and the label A caused by the complex dependence between the features in the detection model, such as a deep network, is eliminated. In this section, we propose mapping traffic features to high-dimensional space via a random Fourier transform to eliminate linear correlations between new features and ensure strict independence between the original features [12], as shown in Eq. (2):

$$H_{RFF} = \{h : f \to \sqrt{2}cos(Gf + \phi) | G \sim N(0, 1), \phi \sim Uniform(0, 2\pi)\} \qquad (2)$$

In Eq. (2), G is sampled from the standard normal distribution, and ϕ is sampled from the uniform distribution in $(0, 2\pi)$. The independence test statistic I_{AB} is defined as the Frobenius norm of the biased cross-covariance matrix:

$$I_{AB} = ||\hat{\Sigma}_{f_A f_B}||_F^2 \qquad (3)$$

The I_{AB} in Eq. (3) is always nonnegative. When I_{AB} decreases to 0, the two variables f_A and f_B tend to be independent. Therefore, I_{AB} can be used to measure the independence between random variables effectively [13].

3.3 Feature Selection Based on Causal Effect

Definition 2: (*do* **Operation**). An important operation method for measuring causal relationships is to change a variable A, keep the generation mechanism of other variables unchanged, and judge the causal relationship between A and F by the change in the resulting variable F[9, 16].

Definition 3: Individual causal effect (ITE). In the potential outcome model, the causal effect between F_i and A can be calculated by changing the value of F_i to obtain different potential outcomes, that is, the individual causal effect of F_i [14]:

$$ITE(AF_i) = Pr\left(F_i^1\right) - Pr(F_i^0) \qquad (4)$$

Definition 4: Average causal effect (ATE). The overall causal effect between A and F_i is defined as the expected value of the ITE, which is called the average causal effect ATE in this paper [14]:

$$ATE(AF_i) = E(ITE(AF_i)) = E\left(F_i^1\right) - E\left(F_i^0\right) \qquad (5)$$

In the SCM constructed in this chapter, the traffic feature dimension corresponding to the network attack is p, and $p >> 1$. If we want to quantitatively analyze the causal relationship between the observed feature set F and cyberattack A, we need to consider the average causal effect between A and F. To better analyze the causal effect between network attacks and features in the sample, the causal direction between A and F is reversed, and the change in A is evaluated by intervening with F. Therefore, according to Eqs. (4) and (5), we define the size of the average causal effect between F and A to judge the causal relationship between A and F.

According to Definition 2, the type of network attack is irrelevant to the noise feature, and the noise feature is related only to the environment. Therefore, when solving for the noise feature, can be simplified as follows:

$$E(F_c, F_n) = E(F_c)E(F_n) \qquad (6)$$

However, in the actual sample data, the causal feature and the noise feature are not independent. Therefore, to ensure that Eq. (6) holds in the real sample, we add weight to the features to make the causal feature and the noise feature independent:

$$E\left(F_c \sum w F_n^T\right) = E\left(F_c W^T\right) E\left(W F_n^T\right) \qquad (7)$$

In Eq. (7), W is a 1-row p-dimensional weight, $\sum w = diag(W_1, W_2, \ldots, W_p)$, and T is the transpose function. However, in the actual sample, we do not know which samples are causal features or noise features.

To accurately obtain the causal feature and noise feature, according to (6) and (7), we use the *do* operation to learn the weight W,

$$W = \arg\min_W \sum_{i=1}^{p} \left\| E\left(F_i^0 \sum w F_{I-i}^0\right) - E\left(F_i^0 W^T\right) E\left(W F_{I-i}^0\right) \right\|_2^2 \quad (8)$$

where I represents all features; i represents the i-th feature; F_i^0 indicates that the i-th feature is treated and set to normal, but the values of the other features remain unchanged; and F_{I-i}^0 indicates that except for the value of the i-th feature remaining unchanged, the rest of the features are treated and set to normal. If we assume that $\sum_{i=1}^{p} W_i = p$, Eq. (8) can be expressed as the loss function $Loss_W$:

$$Loss_W = \sum_{i=1}^{p} \left\| F_i^0 \sum w F_{I-i}^0/p - F_i^0 W^T . W F_{I-i}^0/p^2 \right\|_2^2 \quad (9)$$

Because $E(0) = 1$, if we use W learned from Eq. (9) to assign causal weights to the samples, then

$$E\left(F_c \sum w F_n^T\right) = E\left(F_c W^T\right) \quad (10)$$

If Eq. (10) holds, then $E(A|do(F_n = 1)) - E(A|do(F_n = 0)) = 0$, so F_n is the noise feature that needs to be deleted, and F_c is the causal feature that needs to be selected. Moreover, W is the causal effect between the feature and the attack label.

3.4 Counterfactual Diagnosis

Definition 5 (Causal interpretability). Causal interpretability of the model means that in the anomaly detection model of IoTs, by analyzing the input, output and internal structure of the model, the causal rules of the model detection results are found to explain the model detection results.

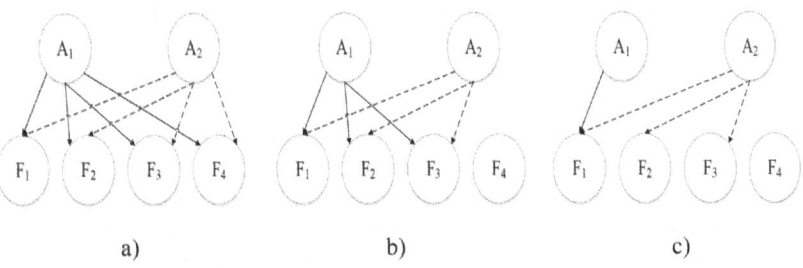

Fig. 2. Relationship graph between network attacks and traffic features

As shown in Fig. 2, the solid line is the "causal link" between A_1 and F, and the dashed line is the "causal link" between A_2 and F. Figure 2a is the SCM between the attack label and the feature shown by the statistical data, Fig. 2b is the SCM after the noise feature is removed from the statistical data, and Fig. 2c is the actual SCM between the attack label and the feature.

After solving stability, the noise feature F_4 was removed, and F_1, F_2 and F_3 were retained; however, when the observed values of the three retained features were all abnormal 1, we could not determine whether it was an A_1 or A_2 attack that caused the abnormality according to Fig. 2b. However, in Fig. 2c, F is clearly caused by an A_2 attack. Therefore, in most cases, after feature deletion, the relationship between the network attack and features is still as shown in Fig. 2b, which is a many-to-many relationship. This makes it difficult for the detection model to distinguish the actual causal relationship between a single attack label and the observed feature vector.

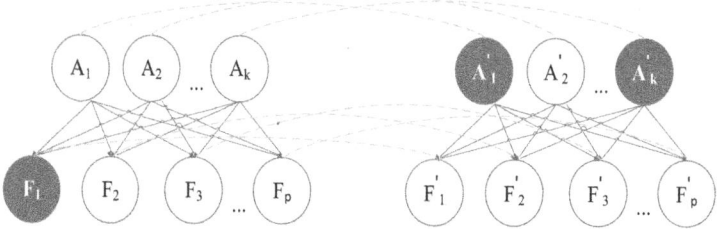

Fig. 3. Dual network SCM diagram

In network attacks, due to the limitations of network reality and other factors, it is impossible to examine and determine which type of attack is most likely to occur through controlled A/B experiments [4]. Moreover, due to the existence of many unobservable or numerous exogenous potential variables u in the network, it is difficult to apply the existing SCM to determine the cause of an attack. Therefore, when applying the formal language of counterfactual reasoning $Pr(\varepsilon = e'|\varepsilon = e, do(A' = a))$ to describe the counterfactual situation of causality between a single network attack and the abnormal feature of test data, converting $Pr(\varepsilon = e'|\varepsilon = e, do(A' = a))$ into an executable mathematical formula by considering the noise factor and variable u at the same time becomes a key task for counterfactual reasoning.

Based on the above problems, in this chapter, a method is proposed the calculates the causal relationship between network attack and feature anomalies in a counterfactual scenario by replicating the double-layer Bayesian network model Fig. 2b into a "dual network" model, performs belief propagation through the "dual network" model, and expresses the causal relationship between network attack and feature anomalies in mathematical language based on existing research [4]. Figure 3 shows the "dual network" model copied from the double-layer Bayesian network model [11]. The left side displays the original factual SCM graph, and the dashed line in the graph represents the exogenous potential variable. The right part of Fig. 3 displays the counterfactual SCM graph, and the variables with apostrophes are counterfactual variables that have the same meaning and assignments as the node variables in the factual graph. After the "dual network" model is constructed, according to Definition 3, operations are performed on the counterfactual graph, and all attacks A' and exogenous potential variables U' except A'_2 are closed and set to 0. In Fig. 3, the blue part on the left is the observed value, and the red part on the right is the intervention value. Since the values of the nonintervened counterfactual nodes are consistent with those of the factual nodes, the nonintervened counterfactual nodes are merged with the factual nodes, and the traffic feature nodes F that are irrelevant to the factual query are deleted, as shown in Fig. 4.

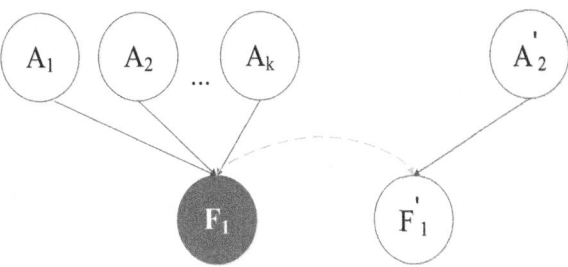

Fig. 4. Counterfactual diagnosis

Assumption 1. In the counterfactual scenario where only the $A = A'$ attack is retained, if the detection data ε still have the most abnormal features, it can be concluded that the A' attack causes the abnormal features of ε, and A' and ε have a causal relationship.

Assumption 2 (Noisy-OR model). In the Noisy-OR model, $Y = f(A_1 \cup A_2 \cup, \ldots \cup A_n)$ is assumed' that is, as long as there is an attack type A_i, the traffic feature F will be abnormal [11].

$$Pr(F = 1|only(A_i = 1)) = 1 \tag{11}$$

Assumption 3. In the noisy-OR model, the probability of any feature F_i being normal ($F_i = 0$) even when it suffers from a network attack ($A_i = 1$) due to the influence of exogenous potential variables and other factors is assumed to be [11].

$$L_{A_i,F} = \prod_{i=1}^{n} Pr\left(F = 0 | A_i = 1, \bigwedge_{j, i \neq j} A_j = 0\right) \tag{12}$$

As shown in the noisy-OR model of Assumption 2, any single attack $A = 1$ is enough to cause the associated traffic anomaly $F = 1$. All the exogenous potential variables and noise variables meet the definition of SCM. Moreover, the causal relationship between an attack and a feature in a counterfactual scenario is equivalent to that in a factual situation. Therefore, assuming that in the case of only one attack, A and ε have a causal relationship and cause the most feature anomalies in ε; then, in the counterfactual scenario, A' and ε also have a causal relationship and also cause the most feature anomalies in ε.

After causal feature selection, in the counterfactual scenario where only attack $A_k = 1$ is retained, the expected number of abnormal features that will still appear [4, 11] is:

$$Es(A_k, \varepsilon) = \sum_{F'} |F'_+| P(F'|\varepsilon, do(Pa(F_+)\backslash A_k = 0))P(F'_+, A_k|\varepsilon) \tag{13}$$

According to Definition 5, in the counterfactual scenario, if there is a causal relationship between A_k and ε, then the expected value of abnormal features $Es(A_k, \varepsilon)$ will be the largest when only $A_k = 1$ is retained. In Eq. (13), A_k represents the k-th attack type, F_+ represents the abnormal feature, F' represents the feature in the counterfactual network, and $\varepsilon = \{F_1 = f_1, F_2 = f_2, \ldots, F_p = f_p\}$ represents the observed feature vector value. $Pa(F_+)\backslash A_k$ represents the set of all parents with counterfactual abnormal evidence feature F_+ except A_k, and $do(Pa(F_+)\backslash A_k = 0)$ represents the counterfactual intervention setting $Pa(F_+)\backslash A_k \to 0$. In Eq. (13), for the observed feature vector F and its counterfactual F', the joint conditional distribution after causal intervention by $do(Pa(F_+)\backslash A_k = 0)$ and $do(U' = 0)$ is [11]:

$$Pr\left(F, F' | \prod_{i=1}^{N} a_i, do\left(\prod_{i \neq k} A_i = 0\right), do(u_k = 0)\right) =$$

$$\begin{cases} Pr(F = 0| \prod_{i=1}^{N} a) & F = 0, F' = 0 \\ 0 & F = 0, F' = 1 \\ L_{A_k,F}^{a_k} Pr(F^{\backslash k} = 1| \prod_{i \neq k} a_i, A_i = 1) & F = 1, F' = 0 \\ \left(1 - L_{A_k,F}\right)\delta(a_k - 1) & F = 1, F' = 1 \end{cases} \tag{14}$$

In Eq. (14), if the observed features are all normal values of 0, it is impossible to obtain abnormal feature values in the counterfactual scenario after all the other network attacks are closed; therefore, the probabilities of $F = 0$ and $F' = 1$ are 0. $F = 0$ and $F' = 0$ represent the probability of no network attack and the probability of normal network traffic when all network attacks participate in the attack. If $F = 1$ and $F' = 0$, the abnormal feature disappears after all other network attacks are closed in the counterfactual scenario; then, the abnormality of the feature is caused by other closed network attacks. Moreover, when $A_k = 1$, the probability of a normal network feature being included is $L_{A_k,F}$. If $F = 1$ and $F' = 1$, then closing other network attacks does not cause abnormal features to disappear; then, the probability of $F = 1$ is equal to 1 minus $L_{A_k,F}$. $\delta(A_k - 1) = 1$ means that this value is equal to 1 if and only if the k-th network attack occurs and 0 otherwise.

4 Detection Performance Evaluation

4.1 Experimental Setup

The real qualities of the CICIDS2019 [4, 15], ToN_IoT [9, 10] datasets were evaluated by using a fuzzy logic system, and the literature [16] shows that the two datasets all contain a group of network attacks that reflect real-world standards and that their generation process fully considers the characteristics of network attacks and the dynamics of network structure.

Table 1. Software and hardware specifications

Hardware Specifications	Software specification
GPU: 3	Operating system: SCMP
CPU: 8	Programming language: Python 3.8
Memory: 16.00 G	Development tools: jupyter
	Language packs: pandas, numpy, sklearn

To use various algorithms more effectively, Python is used to implement network attack detection based on SCM. The hardware and software specifications are shown in Table 1.

4.2 Experimental Results and Analysis

In this section, four experiments performed to verify the performance of the ADCIM method in eliminating redundant information, ensuring stability and ensuring causal interpretability are described.

4.2.1 FSBCE Detection Performance Analysis

a) Comparison of detection accuracy between FSBCE and other feature selection methods

To verify that FSBCE removes noise features and retains causal features, Table 2 below compares the performances of feature reduction methods such as the RF detection method without feature selection, ECSO [17]+RF and causal intervention (CIT) [5]+RF, and PCA [18]+RF, Simplenet [19]+RF, Transformer [20]+RF, CNN [21]+RF and FSBCE+RF of this chapter on two major datasets.

Table 2 shows that, in the two major datasets, compared with RF without dimensionality reduction or feature selection, excluding Transformer's partial improvement in F1 in CICIDS2019, the other methods, such as ECSO, CIT, PCA, Simplenet and CNN, showed a downward trend in F1 after feature reduction. However, FSBCE has the highest F1 on the two major datasets, and its improvement is greater than that of RF without feature selection. Compared with the best previous method, FSBCE improved by >1.2% in the CICDS, >4.7% in the ToN_IoT.

Table 2. Performance of different detection algorithms after feature selection on various datasets.

Detection method	CICIDS2019		ToN_IoT	
	Feature length	F1	Feature length	F1
RF	68	**98.2%**	25	**86.9%**
GWO	39	96.9%	8	86.7%
CIT	38	97.0%	6	79.1%
PCA	51	97.8%	21	83.5%
Simplenet	49	97.7%	23	81.1%
Transformer	49	98.3%	22	82.1%
CNN	51	89.2%	21	80.2%
FSBCE	32	**99.4%**	19	**91.6%**

In summary, after the original features are independently identified via Fourier feature transformation and weighted by FSBCE after training, the noise features are removed to avoid false associations with the attack labels, and the causal features that have a true causal relationship with the attack labels are retained, which significantly improves the detection performance.

b) Performance analysis of FWSCI applied to other detection methods

In addition to comparing the detection performance of several feature selection or dimensionality reduction methods, in this section, FSBCE is also applied to KNN [22], decision tree (DCT) [23], XgBoost [24], logistic regression (LR) [25], SVM [18], CNN [6], GRU [18], LSTM [27], recurrent neural network (RNN) [9], reinforcement learning (RL) [29], deep learning (DL) [8] and other recent methods. As shown in Table 3, "old" denotes the original detection method or model, and "our" denotes that FSBCE was added before the original method training.

As shown in Table 3, after removing the noise features via FSBCE, the detection accuracy of both the machine learning-based and the deep learning-based detection methods improve the two datasets.

Among them, CICIDS2019 had the most obvious effect on DL, with F1 increasing by approximately 8%, while XgBoost had the least obvious effect, with F1 decreasing by less than 0.5%. On the ToN_IoT dataset, after removing the noise features via FSBCE, the effect on DL is also the most obvious, with F1 increasing by approximately 27%. In summary, after removing the noise features via causal inference, FSBCE can effectively improve the detection performances of previous detection methods.

4.2.2 Stability Analysis of Different Network Anomaly Detection Methods

Definition 6 (Stability Performance). Given a training dataset $\{F, A | A = \oint(F)\}$, the task is to learn a detection model to detect whether the network in an unknown environment e (F^e and F are N-IIDs, F^e = F + rand.noise) is attacked, and the detection set F^e has a high minimum accuracy and a small stability error.

Table 3. Detection precision of different detection methods after data processing by FSBCE

Algorithm	CICIDS2019		ToN_IoT	
	old	FSBCE	old	FSBCE
KNN	88.79%	93.67%	86.78%	91.67%
DCT	96.89%	98.89%	87.15%	95.89%
Xgboost	98.22%	98.69%	88.31%	95.69%
LR	96.84%	99.39%	91.69%	97.39%
SVM	98.98%	99.60%	88.40%	96.81%
CNN	89.18%	91.12%	77.72%	91.12%
GRU	93.76%	96.46%	84.78%	96.46%
LSTM	94.87%	96.82%	86.46%	96.82%
RNN	96.73%	98.02%	88.93%	97.02%
RL	97.50%	98.30%	89.86%	97.30%
DL	87.00%	94.93%	77.94%	94.93%

To compare the stability of the proposed method with that of other classic machine learning methods, random noise is added to each feature, starting with the first feature, according to Formula (15) for all the datasets.

$$X'_{ij} = X^*_{ij} + random.normal(0, \sigma) \qquad (15)$$

$random.normal(0, 1)$ is a random noise that follows a standard normal distribution with mean 0 and variance σ.

$$Stb_{Err} = \sqrt{\sum_{j}^{p}(F1_j - F1_{Mean})^2/(p-1)} \qquad (16)$$

$F1_j$ is the detection F1 score after j-dimensional features are contaminated, and $F1_{Mean}$ is the average detection F1 score after random noise is added to 1- to p-dimensional traffic features.

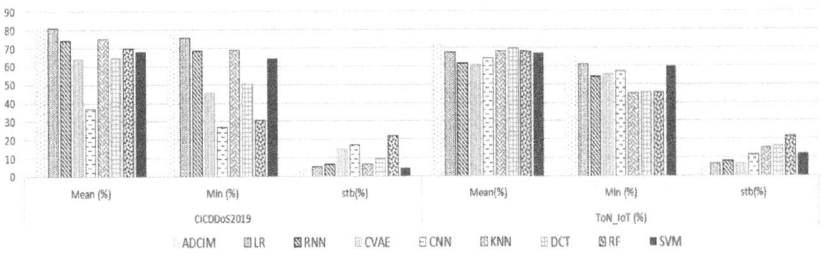

Fig. 5. The average F1 scores and variances of different methods on different datasets

This experiment verifies and compares the stability performance of 9 methods, namely, ADCIM, LR, RNN, conditional variational autoencoder, CVAE [30], CNN [6], KNN [22], DCT [23], RF [24], and SVM [17], on two datasets: CICIDS2019 and ToN_IoT.

Figure 5 shows the average and lowest F1 scores and Stb_{Err} of the 9 methods when $\sigma = 1$. Classic machine learning methods cannot solve the stable prediction problem in all situations. Because they cannot remove the false correlations between noise features and labels in the detection training process, they are often affected by noise features, which leads to unstable detection performance in the same detection dataset. For example, the RF method reaches more than 98% in ToN_IoT when there is no feature pollution, but when there is more feature pollution, its F1 score is as low as approximately 66%. According to Fig. 5, traditional machine learning methods may have better stability in some datasets; however, in other datasets, their F1 scores and stability are often worse, and they lack generalizability.

5 Conclusion

Although machine learning plays a significant role in promoting the development of anomaly traffic detection, it is difficult to achieve stable detection when methods cannot decouple correlation and causality. For this reason, the ADCIM is proposed in this paper. The FSBCE algorithm proposed in this chapter is verified on the CICIDS19 and ToN_IoT datasets, proving that after the FSBCE algorithm processes the data, the number of features needed for training is greatly reduced, the detection accuracy of the training is ensured, and relatively acceptable stability is achieved. Compared with the best previous method, FSBCE improved by >1.2% in the CICDS, >4.7% in the ToN_IoT. Moreover, after verification on two major datasets, compared with that of other optimal methods, the stability of the ADCIM algorithm proposed in this chapter is the best, and its stable error basically remains at a relatively stable interval between 0.038 and 0.062, with the smallest fluctuation among all the other methods. Moreover, the detection accuracy increases by >19%, guaranteeing detection stability when traffic is polluted and causal interpretability for network anomalies.

Acknowledgments. This work is supported by the National Key Research and Development Program of China (No. 2022YFB2901204).

References

1. Moustafa, N., Koroniotis, N., Keshk, M., Zomaya, A.Y., Tari, Z.: Explainable intrusion detection for cyber defenses in the Internet of Things: opportunities and solutions. IEEE Commun. Surv. Tutorials **25**, 1775–1807 (2023)
2. Demertzi, V., Demertzis, S., Demertzis, K.: An overview of cyber threats, attacks and countermeasures on the primary domains of smart cities. Appl. Sci. **13**(2), 790 (2023)
3. Bhaskara, S., Rathore, S.S.: Causal effect analysis-based intrusion detection system for IoT applications. Int. J. Inf. Secur. **22**, 1–16 (2023)

4. Cui, P., Athey, S.: Stable learning establishes some common ground between causal inference and machine learning. Nat. Mach. Intell. **4**(2), 110–115 (2022)
5. Zeng, Z., Peng, W., Zeng, D.: Improving the stability of intrusion detection with causal deep learning. IEEE Trans. Netw. Serv. Manage. **19**(4), 4750–4763 (2022)
6. Thiruloga, S.V., Kukkala, V.K., Pasricha, S.: TENET: temporal CNN with attention for anomaly detection in automotive cyber-physical systems. In: 2022 27th Asia and South Pacific Design Automation Conference (ASP-DAC), pp. 326–331. IEEE, (2022)
7. Dwivedi, A.K.: Anomaly detection in intravehicle networks. ar**v preprint ar**v:2205.03537 (2022)
8. Meyer, P., Hackel, T., Reider, S., Korf, F.: Network anomaly detection in cars: a case for time-sensitive stream filtering and policing. ar**v preprint ar**v:2112.11109 (2021)
9. Sun, H., Chen, M., Weng, J., Liu, Z., Geng, G.: Anomaly detection for in-vehicle network using CNN-LSTM with attention mechanism. IEEE Trans. Veh. Technol. **70**(10), 10880–10893 (2021)
10. Van Wyk, F., Wang, Y., Khojandi, A., Masoud, N., et al.: Real-time sensor anomaly detection and identification in automated vehicles. IEEE Trans. Intell. Transp. Syst. **21**(3), 1264–1276 (2019)
11. Zeng, Z.R., Peng, W., Zeng, D., Zeng, C.: Intrusion detection framework based on causal reasoning for DDoS. J. Inf. Secur. Appl. **65**, 103124 (2022)
12. Zhang, X., Cui, P., Xu, R., Zhou, L., He, Y., Shen, Z.: Deep stable learning for out-of-distribution generalization. In: Proceedings of the IEEE/CVF Conference on Computer Vision and Pattern Recognition, pp. 5372–5382 (2021)
13. Strobl, E.V., Zhang, K., Visweswaran, S.: Approximate Kernel-based conditional independence tests for fast nonparametric causal discovery. J. Causal Inference **7**(1), 20180017 (2019)
14. Rubin, D.B.: Estimating causal effects of treatments in randomized and nonrandomized studies. J. Educ. Psychol. **66**(5), 688–701 (1974)
15. Zeng, Z., Peng, W., Zhao, B.: Improving the accuracy of network intrusion detection with causal machine learning. Secur. Commun. Netw. **2021**, 1–18 (2021)
16. Prasad, M., Tripathi, S., Dahal, K.: An efficient feature selection based Bayesian and rough set approach for intrusion detection. Appl. Soft Comput. **87**, 105980 (2020)
17. Alohali, M.A., Elsadig, M., Al-Wesabi, F.N., Al Duhayyim, M., Hilal, A.M., Motwakel, A.: Swarm intelligence for IoT attack detection in fog-enabled cyber-physical system. Comput. Electr. Eng. **108**, 108676 (2023)
18. Zhou, K., Wang, W., Wu, C., Hu, T.: Practical evaluation of encrypted traffic classification based on a combined method of entropy estimation and neural networks. ETRI J. **42**(3), 311–323 (2020)
19. Liu, Z., Zhou, Y., Xu, Y., Wang, Z.: SimpleNet: a simple network for image anomaly detection and localization. In: Proceedings of the IEEE/CVF Conference on Computer Vision and Pattern Recognition, pp. 20402–20411 (2023)
20. Xu, J., Wu, H., Wang, J., Long, M.: Anomaly transformer: time series anomaly detection with association discrepancy. arXiv preprint arXiv:2110.02642 (2021)
21. Moizuddin, M.D., Jose, M.V.: A bio-inspired hybrid deep learning model for network intrusion detection. Knowl.-Based Syst. **238**, 107894 (2022)
22. Li, W., Yi, P., Wu, Y., Pan, L.: A new intrusion detection system based on KNN classification algorithm in wireless sensor network. J. Electr. Comput. Eng. **2014**, 1–8 (2014)
23. Umar, M.A., Zhanfang, C., Liu, Y.: A hybrid intrusion detection with decision tree for feature selection. arXiv preprint arXiv:2009.13067 (2020)
24. Raghunath, K.M.K., Kumar, V.V., Venkatesan, M., Singh, K.K., Mahesh, T.R., Singh, A.: XGBoost Regression Classifier (XRC) model for cyber attack detection and classification using Inception V4. J. Web Eng. **2022**, 1295–1322 (2022)

25. Nanda, W.D., Sumadi, F.D.S.: LRDDoS attack detection on SD-IoT using random forest with logistic regression coefficient. Jurnal RESTI (Rekayasa Sistem dan Teknologi Informasi) **6**(2), 220–226 (2022)
26. Ma, H., Cao, J., Mi, B., Huang, D., Liu, Y., Li, S.: A GRU-based lightweight system for CAN intrusion detection in real time. Secur. Commun. Netw. **2022**, 1–11 (2022)
27. Shi, Z., Mamun, A.A., Kan, C., Tian, W., Liu, C.: An LSTM-autoencoder based online side channel monitoring approach for cyber-physical attack detection in additive manufacturing. J. Intell. Manuf. 1–17 (2022)
28. Yousuf, O., Mir, R.N.: DDoS attack detection in Internet of Things using recurrent neural network. Comput. Electr. Eng. **101**, 108034 (2022)
29. Ortega-Fernandez, I., Liberati, F.: A review of denial of service attack and mitigation in the smart grid using reinforcement learning. Energies **16**(2), 635 (2023)
30. Yang, Y., Zheng, K., Wu, C., Yang, Y.: Improving the classification effectiveness of intrusion detection by using improved conditional variational autoencoder and deep neural network. Sensors **19**(11), 2528 (2019)

Multi-class Intrusion Detection System in SDN Based on Hybrid LSTM Model

Jue Chen[✉] and Meng Cui

School of Electrical and Electronic Engineering, Shanghai University of Engineering Science, Shanghai 201620, China
jadeschen@sues.edu.cn

Abstract. Software-defined networking (SDN) is a new network paradigm, which is highly decoupled compared to traditional networks. It makes the network innovation easier to operate by separating the data and control planes of the network. However, there are more risks due to the structure of SDN. The attack on the controller will lead to the paralysis of the entire network, so the intrusion detection is particularly important. With the rise of this artificial intelligence, machine learning technology and deep learning technology have been applied in all aspects of life. Deep learning has the advantages of high accuracy, light weight, and fast response speed. Therefore, deep learning technology has also been applied in the field of intrusion detection, while the methods proposed at this stage are mainly concentrated in traditional networks, and are often used to detect Distributed Denial of Service (DDoS) attacks, which lack standardization in attack detection. In this work, we propose a hybrid Long Short Term Memory (LSTM)-based multi-class intrusion detection method to detect 8 common intrusion types on the InSDN dataset. Firstly, a feature selection method is proposed for the high-dimensional data of SDN network, to extract the positive features that are effective for model decision-making, reduce the misleading of the model by unfavorable and negative features, and reduce the computational cost. Secondly, a multi-class intrusion detection model based on multi-output nodes and hybrid LSTM is proposed to improve the accuracy of the model. Finally, this paper compares the proposed model with the machine learning models to verify the performance of the model. Experiments show that the method proposed in this paper improves the accuracy by 37.95%, 25.56%, 23.06% and 10.17% when compared with LOGISTIC, ADABOOST, Random Forest and NETWORK models, respectively, and provides an effective method for multi-class intrusion detection.

Keywords: Software Defined Network · Deep learning · Machine Learning · Multi-class intrusion detection · hybrid LSTM

1 Introduction

Software-Defined Networking (SDN) is an emerging paradigm that promises to change the limitations of traditional networks, by breaking vertical integration, separating the network's control logic from the underlying routers and switches [1], promoting logical centralization of network control, and introducing the ability to program the network. Because of its dynamic, controllable, easy-to-implement economics and adaptable character, SDN is an increasingly popular new technology for today's dynamic systems [2]. Figure 1 depicts the SDN architecture in its entirety, with each layer clearly labeled.

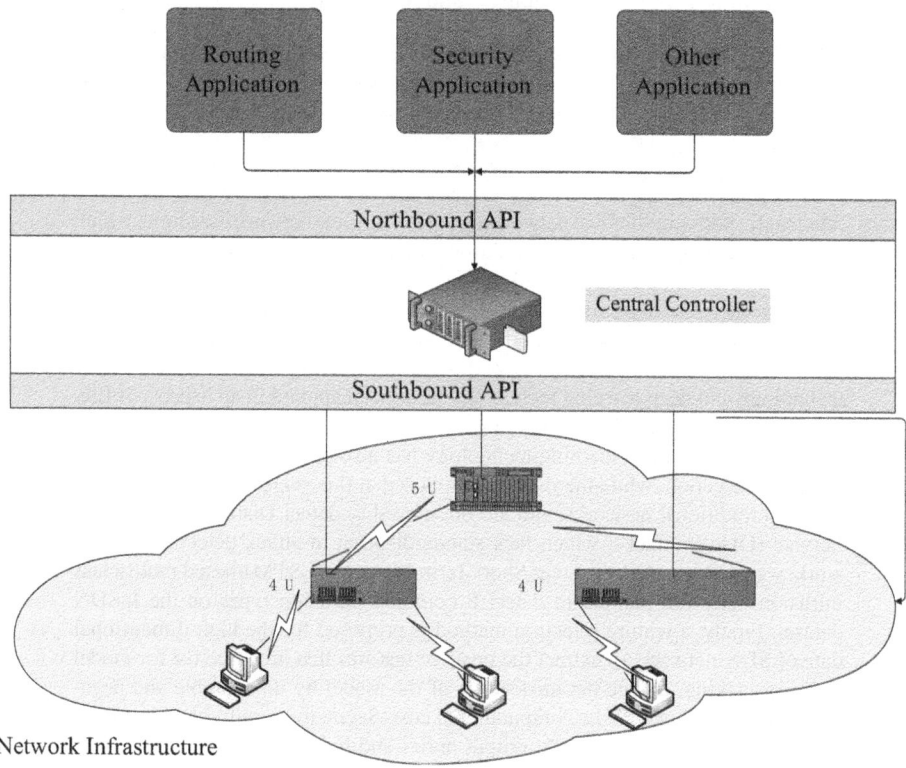

Fig. 1. SDN network structure diagram

Compared with traditional network, the flexible nature of SDN accelerates innovation research and enhances security measures such as threat detection and prevention. Even so, SDN still has security challenges that need to be addressed for broader adoption of the new paradigm. For example, the security breach in the conventional networks has limited damage for only a small part of the network, i.e. probably for the same network vendor, while any attack against the SDN controller may lead to severe consequences on the whole network. In case the attacker succeeds in bringing the controller down, the network might be exposed to severe crashes. With the development of the network and the increase of network applications, users are also paying more and more attention to the privacy of their own networks. Intrusion detection technology has been proved to be effective in blocking network attacks. However, with the continuous development of network attack technology, further development in camouflage requires the intrusion detection system to further adjust to the diversity of attacks and the concealment of attacks. Considering that the network attack has small difference from the normal network, and fast update speed, existing heuristic algorithms and firewalls can no longer meet the needs of existing intrusion detection. Thanks to the continuous development of artificial intelligence technology, the machine learning methods have proved their effectiveness in traffic prediction, especially in the filed of computer networks. Different probability theories and statistics-based behaviors with sufficient data sets can be used to identify network anomalies.

Dishan Jing [3–5] proposed multi-classification methods based on support vector machine, knn and Naive Bayes, respectively, and the classification accuracy rates reached 84%, 84.82% and 87%, respectively. Although these machine learning-based intrusion detection techniques achieve good results, high computational costs are inevitable. Moreover, the learning of data sets is not deep enough, which leads to frequent false alarms. Compared with traditional machine learning, deep learning shows higher performance in intrusion detection, because of its ability to extract feature values and process data in a more intelligent and automatic way. Chuanlong Yin et al. [6] proposed an Recurrent Neural Network (RNN)-based intrusion detection system that achieved an accuracy of 81.29% in the five-category test on the KDD dataset. Tuan A Tang et al. [7] proposed an intrusion detection method using DNN in SDN, and the accuracy rate reached 75.5% in the case of a small number of eigenvalues. Sara A. Althubiti et al. [8] proposed to use Long Short-Term Memory (LSTM) for intrusion detection, and achieve 85% accuracy on the CIDDS data set.

It can be inferred that deep learning has been widely used in the field of intrusion detection, but is often limited to applying for traditional networks, and does not consider SDN scenarios. Moreover, the binary classification methods are most widely used which can only distinguish normal traffic from abnormal traffic. With the increase of network intrusion types, the requirements for network intrusion detection increase as well, and intrusion detection models need to be able to detect and identify attacks more accurately and quickly. Therefore, a multi-classification method is more in line with the requirements of the present stage. In this paper, we propose a multi-class intrusion detection method based on GA-CNN-BILSTM, and choose the SDN dataset InSDN to enhance the accuracy of the intrusion detection for SDN attacks.

2 Related Work

Even though SDN design has many advantages, it still has to face challenges including attacks, vulnerabilities, and threat vectors [9]. Currently, SDN security is a very important topic in academic research, as SDN networks can be targeted by attackers through an efficient way [10]. (It has become the support for reducing cyberspace penetration attacks). Therefore, a network intrusion detection system is necessary to combat with network threats. However, due to the particularity of the SDN network, its data is more complex and data processing sets a higher demand. When being faced with a large number of data and feature values, we need to filter to reduce the redundancy.

Binh Tran et al. [11] proposed a particle swarm optimization algorithm (PSO) for high-dimensional data, which can effectively reduce data dimension, and significantly improve the classification accuracy. Kasongo and Sun [12] proposed an intrusion detection system based on Deep Learning (DL). They combined a filter-oriented feature selection with feed-forward DNN to detect network intruders. The approach utilized an information gain mechanism and outperformed most existing traditional ML approaches through experiments. Wenping Ma et al. [13] proposed an ant colony optimization algorithm for high-dimensional feature selection, which has state-of-the-art performance on most datasets. Excellent feature selection methods can help us better process data, and can also further improve the accuracy of model detection. Mazini et al. [14] developed a hybrid Artificial Bee Colony (ABC) and Ada Boost algorithms for anomaly detection in NSL-KDD and ISCXIDS2012 datasets with 98.9% detection accuracy, and 1.1% false-positive rate. Khraisat et al. [15] proposed an ensemble of hybrid intrusion detection system Intrusion Detection System (IDS) through combining C5 classifier with One-Class SVM classifier. This hybrid model was tested on the Bot-IoT dataset, which achieved 94% malware detection and 99.97% overall accuracy.

YakubuImrana et al. [16] proposed a BiDLSTM deep learning model. Authors used the first LSTM on the original input data and the other on a reversed replica of the input data, which solved the problem of gradient disappearance in RNN and improved the performance of the model on classification problems. JingmeiLiu et al. [17] proposed a fast intrusion detection method based on LightGBM, which normalized the data, increased minority samples through ADASYN oversampling technology, and finally used the LightGBM ensemble learning model to reduce time complexity while ensuring accuracy. Mahmoud et al. [18] proposed a new regularization method using SD-Reg to solve the overfitting problem. At the same time, CNN can be effectively combined into the lightweight intrusion detection systems since it solves the parameter explosion problem of traditional neural networks. Devan and Khare [19] presented a deep neural network (DNN) based classification approach for IDS. Initially, they perform normalization, implemented XGBoost technique for feature selection, and employed DNN for classification process. During the training process of DNN, the Adam optimizer is used to maximize learning rate. Fan Jiajia et al. [20] proposed an intrusion detection model based on SAE-BALSTM. In this model, the feature selection was based on SAE, which performed better than principal component analysis (PCA) in the aspect of dimensionality reduction effect, and solved the overfitting problem. The BALSTM model was composed of forward ALSTM and backward ALSTM connections, which can automatically learn important features. Moreover, to prevent over-fitting, a dropout layer is added behind the BALSTM layer to randomly deactivate the neurons of BALSTM model.

3 Methodology

In this section, we first analyze the differences between traditional network datasets and SDN datasets, and decide to use the latter one. Then, we explain the genetic algorithm used for feature selection architecture and further introduce the architecture of the proposed hybrid LSTM network.

a) Datasets

With the emergence of the SDN network paradigm, IDS with a centralized view has become possible, but the performance of these detection techniques depends on the quality of the training data set. In previous related works, researchers often used data sets such as KDD CUP99, NSL-KDD and so on, which are generated from traditional networks. On one hand, the KDD Cup 99 dataset is a dataset for a network intrusion detection competition presented at the International Conference on Knowledge Discovery and Data Mining in 1999. Different types of attacks such as Denial of Service (DoS) attacks, remote to local (R2L) attacks are simulated on a real network environment named "Georgia Tech Network Simulator" (GTNetS) through DND (DARPA Intrusion Detection Evaluation Program). In brief, this dataset contains a total of 22 different types of attacks and 41 eigenvalues. On the other hand, the NSL-KDD dataset is an improved version of the KDD-CUP 99, which solves the limitations of the original dataset, such as the existence of duplicate and redundant records, and the lack of diversity in attack types. The data of the NSL-KDD is collected through LAN emulators containing mawi and Tstat tools, and the mawi and Tstat are responsible for collecting data and generating network attack traffic, respectively. Moreover, the dataset contains five types of attacks: DOS, R2L, U2R, Probe and Normal, and 42 signatures. Among them, the first four classes represent different types of attacks, while the Normal class represents normal network traffic. Different from the above two datasets, the InSDN dataset [21] dataset is a public SDN dataset released by researchers, and can be directly applied into the SDN network anomaly detection system. The comparison among these three datasets is shown in Table 1.

InSDN uses Mininet for data collection, chooses RYU as the controller, and adopts Kali Linux to perform various attack scenarios and create different attack classes such as DoS, DDoS, Web attacks, Password-Guessing, Botnet, Exploitation, and Probe attacks. The dataset includes 83 traffic characteristics, and all the data is marked with the attack type. In order to generate the dataset, four VMware virtual machines are used for network attack. The attack types and tools are shown in Table 2. In addition, This article is aimed at the SDN scenario, so the InSDN dataset is used. Since the InSDNdataset has a large number of redundant features and data, it requires a feature selection process before IDS training.

Table 1. Comparison of Datasets

Name	Environment	build tool	Number of attack types	number of features
KDD-Cup99	GTNetS	DND, DARPA	Four	41
NSL-KDD	local area network	Mawi, TstatD	Five	42
INSDN	MININET	Hping3	Eight	83

b) Feature selection

In deep learning models, dimensionality reduction plays an extremely major role in reducing the number of redundant attributes considered, thereby reducing the time complexity by selecting the most significant attributes contributing towards improvement of prediction result. In the model training phase, if the number of features is found to be

Table 2. InSDNDataset

Attack Classes	Description of Activities	Attack Tools	Attacker Machine	Victim-Network
DoS	TCP-ACK flood, UDP flood, HTTP flood, slow-rate, HTTP POST, Slowloris	LOIC, slowhttptest, HULK, torshammer	KaliLinux: 200.175.2.130	Vhost(h4): 192.168.20.134
	Slowloris, TCP flood	Nping, Metasploit framework	Kali-Linux: 200.175.2.130	Metasploitable2: 192.168.3.130
DDoS	TCP-SYN Flood, UDP Flood, ICMP Flood	Hping3	h1:192.168.20.131 h2:192.168.20.132	h4:192.168.20.134 Metasploitable2: 192.168.3.130
Web Attacks	XXS, Sql Inject	Metasploit framework, sqlmap	Kali-Linux: 200.175.2.130	Web server(DVWA): 172.17.0.1
R2L	Password-Guessing Attack	Burp Suite, hydra	Kali-Linux: 200.175.2.130	Web server (DVWA): 172.17.0.1
		Metasploit framework	Kali-Linux: 200.175.2.130	Metasploitable2: 192.168.3.130
Malware	Botnet attack	ARES	Kali-Linux 200.175.2.130	h1:192.168.20.131 h2:192.168.20.132
Probe	version scan, Port Scan, discover services	Nmap	Kali-Linux: 200.175.2.130	h1:192.168.20.131 h2:192.168.20.132 h3:192.168.20.133 h4:192.168.20.134
	Port Scan, vulnerability scan (WMAP)	Metasploit framework	Kali-Linux: 200.175.2.130	Metasploitable2: 192.168.3.130
U2R (Exploitation)	Vsftpd, IRCd, Samba and distcc	Metasploit framework	Kali-Linux: 200.175.2.130	Metasploitable 2:192.168.3.130

high, the over-fitting phenomenon may be caused. In this paper, Genetic algorithm (GA) is exploited to handle this issue and is used for dimensionality reduction. The search for the optimal feature subset is a process of genetic variation, and the optimal solution is achieved through iterations until the convergence is reached. As mentioned before, the InSDN dataset is composed of *83* features. As a result, the Chromosome encodes an *83*-dimensional row vector and sets the expected dimension to be *20*.

c) Proposed network model

This section describes the model architecture proposed in this paper in detail. The model includes input embedding layer, hybrid BiLSTM layer, attention mechanism layer, and classification layer (i.e., CNN layers). The overall structure of proposed model is shown in Fig. 2. The component structure of model will be described in detail below.

BiLSTM

The bidirectional LSTM augments standard LSTMs to improve a model's performance on classification issues. It trains two LSTMs on the input data. The first LSTM is trained on the original input data and the other on a reversed replica of the input data. By

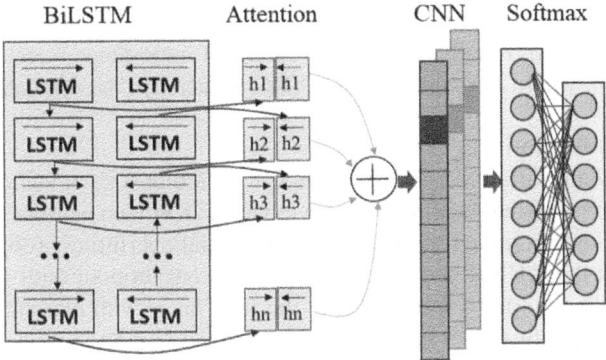

Fig. 2. Proposed hybrid CNNA-BiLSTM model

this, more meaning is added to the network and faster results are achieved. The concept behind BiLSTM is very straightforward. It comprises duplicating the first recurrent layer in the network, then provides the input data in its original form as input to the first layer, and a reversed replica of the input data to the duplicated layer. This concept solves the problem of vanishing gradient in standard RNNs. Finally, the output of BiLSTM is $h = \{h_1, h_2 \cdots h_n\}$.

$$h_i = \overleftarrow{h_i} || \overrightarrow{h_i} \; h_i \in R^{2d} \tag{1}$$

|| is the operator of concatenation and d *is* the dimension of the LSTM in terms of hidden units.

Attention
Generally, not all features in dataset contribute equally to the representation of intrusion detection, so we leverage feature attention mechanism to capture the distinguished influence of the features on the emotion of intrusion detection, and then form a dense vector considering the weights of different feature vectors. Specifically, we have:

$$u_{ti} = \tanh(Wh_{ti} + b) \tag{2}$$

$$a_{ti} = \frac{\exp(u_{ti}^T u_w)}{\sum_{j=1}^{n} \exp(u_{tj}^T u_w)} \tag{3}$$

$$s_t = \sum_i a_{ti} h_{ti} \tag{4}$$

t represents t-th dataflow, i represents i-th feature in the dataflow and n is the number of factors in a dataflow. h_{ti} represents the factors annotation of the i-th in the t-th dataflow which fed to a one-layer MLP to get h_{ti} as a hidden representation of h_{ti}. More specifically h_{ti} is the concatenation output of the BiLSTM layer in our model. W is a weight matrix of the MLP, and b is a bias vector of the MLP. Then we measure the importance of factors through the similarity between h_{ti} and a word level context vector u_w which is randomly initialized. Finally, we get a normalized importance weight a_{ti} through a softmax function, where a_{ti} is the weight of the i-th factor in the t-th dataflow.

CNN and Classification

The attention layer is followed by CNN layers with different sizes of neurons. The output of attention layer is fed into the first CNN layer with 128 hidden neurons. The activation function of this layer is *tanh*. In order to avoid potential overfitting problem, dropout is utilized between these CNN layers, and we try different dropout rates to find the best configurations. The output is then fed into the next layer with classification, and the activation function in this layer is Softmax.

d) Evaluation metrics

To evaluate our model's performance, we calculate the accuracy, recall, specificity, and false alarm rate, and we also investigate the model's precision and F-score.

Each of these metrics is explained and derived as follows.

(i) Accuracy (ACC): This is the ratio of the number of correctly detected intrusions to the total number of traffic records.

$$ACC = \frac{TP + TN}{TP + TN + FP + FN} \tag{5}$$

(ii) Recall: It refers to the ratio of the number of intrusion records correctly detected as intrusions to the overall anomalies…

$$Recall = \frac{TP}{FN + TP} \tag{6}$$

(iii) Precision: This refers to the ratio of the true anomalous records to the overall traffic records identified as intrusions…

$$Precision = \frac{TP}{TP + FP} \tag{7}$$

(iv) F-Score: It refers to the harmonic mean of the precision and true positive rate…

$$F - Score = 2\left(\frac{1}{Precision^{-1} + TPR^{-1}}\right) \tag{8}$$

4 Experiments and Results

In this section, we present the implementation of GA-CNN-BILSTM and discuss the experimental findings. The various performances of this model are summarized in Table 3. We compare the model's performance with state-of-the-art methods trained and tested on the same dataset (the InSDN dataset). Also, we present a comparison of results with some recently published methods on the InSDNdataset.

a) Experiment setup

Table 3. Summary of the proposed model architecture

Layer	Type	Output shape	Number of parameters
1	Conv1D	10,16	1040
2	BatchNormalization	10,16	64
3	Activation	10,16	0
4	Bidirection	10,64	12544
5	Dropout	10,64	0
6	Attention	128	20480
7	BatchNormalization_1	128	512
8	Activation	128	0
9	Softmax	8	1032

A Multi-class intrusion detection system in SDNs based on GA-CNN-BILSTM model is proposed in this paper. Python programming language is utilized to implement the different phases of the proposed method. To be precise, we use Python's TensorFlow and Keras libraries to implement the various components of the model. All models used in this work are compiled with GPU support, and all experimental studies are conducted on a Conda environment on Ubuntu Desktop 18.04.4 LTS operating system running on Intel(R) Xeon(R) Gold 5118 CPU @2.30 GHz CPU and Nvidia GeForce RTX 2080Ti 11 GB GPU. Comprehensively consider hardware requirements and training effects, training and testing sets are divided into a series of batches of 128, and each batch is processed in sequence during one training epoch. The number of epochs is set to be 50, and the learning rate is set to be 0.0001. Additionally, Adam is used to optimize the network. Its pseudocode is shown in Algorithm 1.

Algorithm 1: Hybrid model Training

1. Load dataset
2. **For** Data in Training and Test Sets **do**
3.   Extract Features (x)
4.   Encode Labels (y)
    Input: Features Extracted
    Output: Classifications
5. **For** Feature in x **do**
6.   **If** Feature = Nonnumerical then
7.   Encode Feature using Keras Library
8. **For** i from 1→n **do**
9.   Start: K=N/Batch
10.   Split Training set into K-groups
11.   Load proposed model
12.   Fit model with K-1 group
13.   Validate model with remaining Kth group
14.   Repeat until all K-groups are used as validation set
15. Test model on Test sets

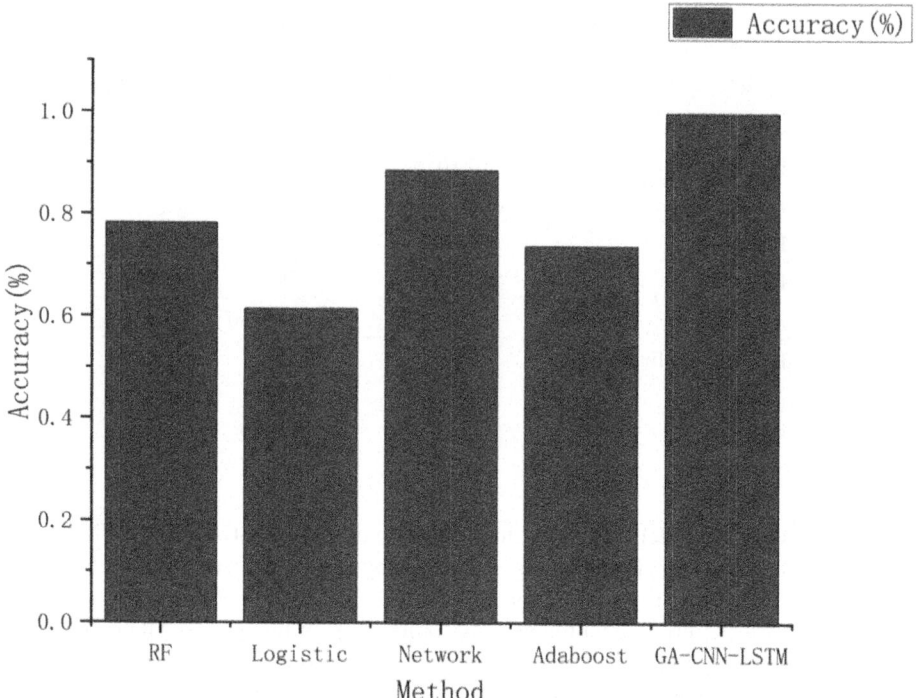

Fig. 3. Comparison of Detection Accuracy among Different Models

b) Performance Comparisons

In this experiment, an eight-level classifier (Normal, DDoS, DoS, Probe, U2R, BFA, BOTNET, WEB) is provided for training all features. Here we compare the proposed GA-CNN-BILSTM model with traditional machine learning models, including Random Forest (RF), LOGISTIC, NETWORK, ADABOOST, The evaluated metrics include precision, recall, F-score and accuracy, and the evaluation results are shown in Table 4 and Fig. 3. The model proposed in this article has better performance than machine learning models in parameters such as recall, F-1 score, accuracy. It can be concluded that the proposed hybrid LSTM model not only improves the performance, but compared with the existing IDS model, the GA-CNN-BILSTM model highlighted in this paper has achieved 99.31% multi-classification performance on the SDN network dataset. Compared with the CNN-LSTM model, our proposed GA-CNN-BILSTM model achieves a better result of 92.43% in recall rate, and 93.56% in F-score. All in all, the proposed GA-CNN-BILSTM shows superiority in intrusion detection compared with existing models.

Table 4. Comprehensive analysis using INSDN

Method	Performance in percentage			
	Precision	Recall	F-Score	Accuracy
RF	82.26%	70.63%	66.24%	78.25%
LOGISTIC	60.35%	56.82%	48.17%	61.36%
NETWORK	79.63%	72.31%	71.24%	88.48%
ADABOOST	78.42%	76.33%	80.29%	73.75%
GA-CNN-BILSTM	94.32%	92.43%	93.56%	99.31%

5 Conclusions

This work proposes a deep learning algorithm for intrusion detection in SDN, the GA-CNN-BILSTM model. This model has good performance and accurate calculation results, and has achieved excellent results on solving multi-classification intrusion detection problems. In order to verify the performance of the proposed model, we use InSDNdataset (the SDN dataset) to tame the model. (The InSDNdataset is currently the most used SDN dataset and is more suitable for the SDN network environment than traditional network datasets.) The experimental results show that the GA-CNN-BILSTM model is superior to the traditional Machine Learning in terms of accuracy, precision, recall and F-score, and the proposed model also improve detection performance when compared with other literature models, Moreover, our proposed model not only improves the overall anomaly detection rate, but also detects eight common types of network traffic which improves the detection rate of each attack category. In the future, we plan to develop and explore more advanced feature selection methods, and change the model structure to further reduce the response time and construct a faster integration platform.

Acknowledgement. This work is supported by the National Natural Science Foundation of China and The Research Project of Shanghai Science and Technology Commission (Grant No. 62102241, No. 23ZR142540).

Declaration of Competing Interest. The authors declare that they have no known competing financial interests or personal relationships that could have appeared to influence the work reported in this paper.

References

1. Van Adrichem, N.L.M., Van Asten, B.J., Kuipers, F.A.: Fast recovery in software-defined networks. In: 2014 Third European Workshop on Software Defined Networks, pp. 61–66. IEEE (2014)
2. Chen, X., Wang, X., Yi, B., He, Q., Huang, M.: Deep learning-based traffic prediction for energy efficiency optimization in software-defined networking. IEEE Syst. J. **15**(4), 5583–5594 (2021). https://doi.org/10.1109/JSYST.2020.3009315

3. Ye, F., Tang, T.-A.: Institute of Electrical and Electronics Engineers. Beijing Section, C. Fu dan da xue (Shanghai, and Institute of Electrical and Electronics Engineers, Proceedings, 2019 IEEE 13th International Conference on ASIC (ASICON 2019), 29 October–1 November 2019, Chongqing, China (2019)
4. Pajouh, H.H., Javidan, R., Khayami, R., Dehghantanha, A., Choo, K.K.R.: A Two-layer dimension reduction and two-tier classification model for anomaly-based intrusion detection in IoT backbone networks. IEEE Trans. Emerg. Top. Comput. **7**(2), 314–323 (2019). https://doi.org/10.1109/TETC.2016.2633228
5. Gumus, F., Sakar, C.O., Erdem, Z., Kursun, O.: Online Naive Bayes classification for network intrusion detection. In: ASONAM 2014 - Proceedings of the 2014 IEEE/ACM International Conference on Advances in Social Networks Analysis and Mining, pp. 670–674, October 2014. https://doi.org/10.1109/ASONAM.2014.6921657
6. Yin, C., Zhu, Y., Fei, J., He, X.: A deep learning approach for intrusion detection using recurrent neural networks. IEEE Access **5**, 21954–21961 (2017). https://doi.org/10.1109/ACCESS.2017.2762418
7. Tang, T.A., Mhamdi, L., McLernon, D., Zaidi, S.A.R., Ghogho, M.: Deep learning approach for network intrusion detection in software defined networking. In: Proceedings - 2016 International Conference on Wireless Networks and Mobile Communications, WINCOM 2016: Green Communications and Networking, Dec. 2016, pp. 258–263 (2016). https://doi.org/10.1109/WINCOM.2016.7777224
8. Althubiti, S.A., Jones, E.M., Roy, K.: LSTM for Anomaly-Based Network Intrusion Detection; LSTM for Anomaly-Based Network Intrusion Detection (2018)
9. Susilo, B., Sari, R.F.: Intrusion detection in software defined network using deep learning approach. In: 2021 IEEE 11th Annual Computing and Communication Workshop and Conference, CCWC 2021, pp. 807–812, January 2021. https://doi.org/10.1109/CCWC51732.2021.9375951
10. Corsini, A., Yang, S.J., Apruzzese, G.: On the evaluation of sequential machine learning for network intrusion detection. In: ACM International Conference Proceeding Series, August 2021. https://doi.org/10.1145/3465481.3470065
11. Tran, B., Xue, B., Zhang, M.: Variable-length particle swarm optimisation for feature selection on high-dimensional classification (2019)
12. Kasongo, S.M., Sun, Y.: A deep learning method with filter based feature engineering for wireless intrusion detection system. IEEE Access **7**, 38597–38607 (2019). https://doi.org/10.1109/ACCESS.2019.2905633
13. Ma, W., Zhou, X., Zhu, H., Li, L., Jiao, L.: A two-stage hybrid ant colony optimization for high-dimensional feature selection. Pattern Recognit. **116**, 107933 (2021). https://doi.org/10.1016/j.patcog.2021.107933
14. Mazini, M., Shirazi, B., Mahdavi, I.: Anomaly network-based intrusion detection system using a reliable hybrid artificial bee colony and AdaBoost algorithms. J. King Saud Univ. – Comput. Inf. Sci. **31**(4), 541–553 (2019). https://doi.org/10.1016/j.jksuci.2018.03.011
15. Khraisat, A., Gondal, I., Vamplew, P., Kamruzzaman, J., Alazab, A.: A novel ensemble of hybrid intrusion detection system for detecting internet of things attacks. Electronics (Switzerland) **8**(11), 1210 (2019). https://doi.org/10.3390/electronics8111210
16. Imrana, Y., Xiang, Y., Ali, L., Abdul-Rauf, Z.: A bidirectional LSTM deep learning approach for intrusion detection. Expert Syst. Appl. **185**, 115524 (2021). https://doi.org/10.1016/j.eswa.2021.115524
17. Liu, J., Gao, Y., Hu, F.: A fast network intrusion detection system using adaptive synthetic oversampling and LightGBM. Comput. Secur. **106**, 102289 (2021). https://doi.org/10.1016/j.cose.2021.102289

18. ElSayed, M.S., Le-Khac, N.A., Albahar, M.A., Jurcut, A.: A novel hybrid model for intrusion detection systems in SDNs based on CNN and a new regularization technique. J. Netw. Comput. Appl. **191**, 103160 (2021). https://doi.org/10.1016/j.jnca.2021.103160
19. Devan, P., Khare, N.: An efficient XGBoost–DNN-based classification model for network intrusion detection system. Neural Comput. Appl. **32**(16), 12499–12514 (2020). https://doi.org/10.1007/s00521-020-04708-x
20. Jiajia, F., Jiangfeng, X., Junfeng, Z.: Intrusion detection model based on SAE and BALSTM. In: 2021 IEEE International Conference on Artificial Intelligence and Computer Applications, ICAICA 2021, pp. 1192–1197, June 2021. https://doi.org/10.1109/ICAICA52286.2021.9498102
21. Elsayed, M.S., Le-Khac, N.A., Jurcut, A.D.: InSDN: a novel SDN intrusion dataset. IEEE Access **8**, 165263–165284 (2020). https://doi.org/10.1109/ACCESS.2020.3022633

Design and Implementation of Computing Based Service Chain Orchestration Framework

Dongsheng Qian[1], Yusheng Lv[1], Kuo Guo[1], Shang Liu[1], Xu Huang[1], Chenxi Liao[1], Jingjing Liu[2], Xiaolong Liu[3], Kai Chen[3], and Jia Chen[1,4](✉)

[1] Beijing Jiaotong University, Beijing, China
chenjia@bjtu.edu.cn
[2] China Mobile Group Liaoning Company Limited, Shenyang, China
[3] China Academy of Information and Communications Technology, Beijing, China
[4] Peng Cheng Laboratory, Shenzhen, China

Abstract. The computing power network intends to accomplish on-demand allocation and flexible scheduling of network computing resources as well as the unification of the administration of multiple computing resources. Two significant obstacles stand in the way of its progress, nevertheless. First off, heterogeneous computing resources like CPU, GPU, and FPGA lack a consistent measure. Second, service function chains created by breaking down individual large-scale computing activities lack an effective orchestration method. This paper designs and implements a computing based service chain framework to effectively utilize diverse computing resources in the network and deliver efficient computing services. The framework has the ability to efficiently handle user computing tasks, assess user computing demands, and standardize the evaluation of computing resources. In addition, we propose a Stackelberg method to deal with the price of resources and task distribution issues between computing service users and providers. When the user scale is less than 30, the algorithm finishes the computation in one minute, showing its effectiveness in handling tasks using the computing service chain. We confirm its efficacy in lowering user costs and raising service provider revenue through a comparison analysis with single-node processing of the same jobs.

Keywords: Computing Based Service Chain · Programmable Data Plane · Gaming Theory · Orchestration

1 Introduction

As 6G technology advances, it will be able to better serve use cases like autonomous driving, holographic communication, and smart cities. However, these application scenarios face a number of significant obstacles, such as heterogeneous data analysis and huge data transmission. The network of computer power has emerged in this background. By fusing networking with pervasive computing, the computing power network seeks to create a new architecture [1]. To meet the needs of network flexibility and Quality of Service (QoS) for enterprises, the computing power network requires dynamic scheduling of jobs to the best computing nodes for processing, achieving interconnection among distributed nodes. In order to accomplish this, the system must organize service function chains while decomposing user computing activities. We propose using the computing service chain to support these services so they may fulfill user requirements through global resource metrics, comprehensive planning, and effective scheduling, which includes the following three factors.

(1) The issue of translating user requirements into equivalent hardware computing resource quantities is dealt with by computing power measurement and modeling. As a result, it is possible to create a single resource description for heterogeneous computing resources and build a connection between user needs and available hardware.
(2) Identification and tracking of computing power addresses challenges with global identification, planning, and administration of computing jobs produced by user requirements. Based on the identification, the system can conduct global tracking operations.
(3) To meet user expectations, computing power orchestration and management address the orchestration and scheduling of service function chains under resource constraints. With this technology, it will be possible to optimize the configuration and scheduling of computing resources, increase computing performance, and guarantee service quality.

Currently, relevant research has provided theoretical and technical foundations. Technologies such as Programming protocol-independent packet processors [2], in-band telemetry [3], network function virtualization [4], and service function chaining [5] have emerged and combined with networking, providing partial technical foundations to address the four core problems in the computing power network. Reference [6] proposed a lightweight service function chaining framework in SINET [7, 8] and implemented virtual network functions using Docker containers. It expanded network functionality but did not delve into other issues. References [9, 10] utilize the separation of data and control plane in SDN to achieve more flexible service support, but they do not explore resource integration and utilization. References [11, 12] provide services for data-intensive applications using edge computing and service function chaining, but lack exploration of various types of computing resources and multi-user usage. Reference [13] studies the resource allocation solution for multi-tenant customized services in network slicing scenarios, but lacks decomposition of tasks and comprehensive resource utilization.

This paper introduces the ground-breaking idea of "Computing Power as a Service" and incorporates it into the network. To be more precise, based on computing power needs, they are split up into various computing tasks and deployed on various computing nodes. A chain of computing power service nodes is created by connecting them. We design a computing service chain orchestration system aiming to provide flexible and dynamic computing services, utilizing heterogeneous computing resources combining computation and networking, and improving the efficiency of computing tasks while saving network resources. The main contributions of this paper are as follows:

1. We propose a Computing based service chain orchestration framework that can enable multi-dimensional computing power measurement and modeling while constituting a single dimension. Additionally, we include the computing power identification method into already in use network protocols, allowing for efficient task scheduling and user computing tasks tracking.
2. We offer a Stackelberg game algorithm that takes into account the profitability of both computing users and computing suppliers, addressing the pricing and job allocation issues in the network. This approach is based on the computing power evaluation models.
3. We test the algorithms based on programmable networks by implementing the aforementioned processes in a prototype and achieving functions such as modeling computing power awareness, orchestrating computing tasks, and scheduling. We demonstrate the advantages of our system in terms of latency and revenue by contrasting the task decomposition strategy chosen by the Stackelberg algorithm with the strategies of equal distribution and single-node execution.

The rest of the paper is organized as follows: Sect. 2 introduces the overall model architecture. Section 3 describes the algorithms. Section 4 presents the experimental system. Section 5 conducts experiments and analysis. Section 6 concludes the paper with a summary.

2 System Design of the C-SFC Framework

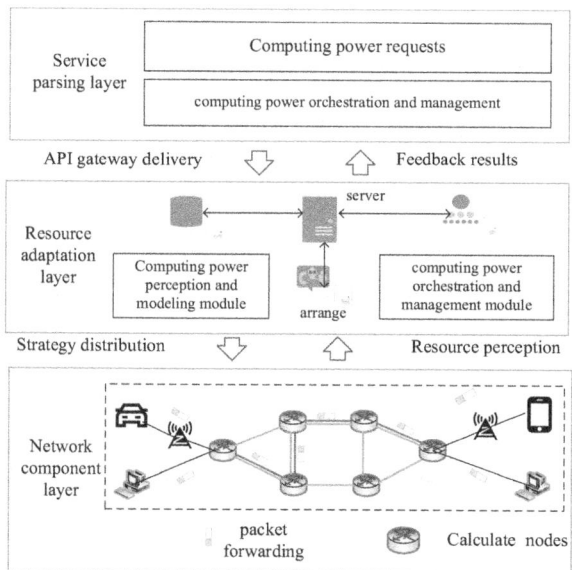

Fig. 1. System overall architecture

The service parsing layer, resource adaption layer, and network component layer are the three levels that make up the overall architecture, as depicted in Fig. 1. The layer that interacts with users, known as the service parsing layer, is in charge of task identification and user demand analysis in order to finish computing jobs. Under resource limitations, the resource adaption layer orchestrates service function chains. In addition to performing multi-dimensional resource awareness on the underlying network and computing nodes, it receives computing power service description files from above. The orchestration strategies for the computing power service chain are then implemented. Computing nodes, containers, and software switches are all part of the network's component layer. The computing power awareness and modeling module, computing power identification and tracking module, and computing power orchestration and management module are the three main functional modules that the system model needs to implement these features. In the diagram below, the three modules are depicted.

To facilitate management, we represent the entire network as a connected graph $G = (N, L)$, where N represents the set of nodes and L represents the set of links. Next, we establish models for the three functional modules.

2.1 Computing Power Awareness and Modeling

Different business scenarios have different requirements for computing resources. As various scenarios and applications involve heterogeneous multidimensional computing resources such as computation, storage, and networking, it is necessary to have a unified heterogeneous computing modeling system to quantify and model the underlying heterogeneous computing resources. In this modeling system, certain statistics need to be performed on device information, including device availability, device network address, device computing resource category, and device computing resource quantity. The device computing resource quantity includes various indicators such as CPU utilization, memory, bandwidth, storage space, floating-point computing capability, and latency estimation, which are recorded as $S_N = <cpu\_util, mem\_ava, eth\_speed, space\_ava, FLOPS, delay\_eva >$. Various data is collected through calls to the Zabbix API and shell scripts.

For floating point computing power, the formula is as follows:

$$flops = CPU\_frequency * CPU\_CORES * op\_unit \qquad (1)$$

The *flops* is the floating-point computing power, CPU_frequency represents the main frequency of the central processing unit, and CPU_CORES is the number of CPU cores, depending on the specific server. op_unit is the number of floating-point computing units, related to the architecture.

Delay evaluation represents the average delay between each node and all its directly connected nodes. The calculation formula for the delay evaluation between computing node I and the other j nodes is as follows:

$$delay\_eva = \frac{\sum_j delay_{i->j}}{\sum_j 1} \qquad (2)$$

This paper adopts the Multi-attribute Decision Making [14] method as the modeling approach, which consists of two key elements. Firstly, it involves obtaining attribute values and attribute weights through a certain method. Secondly, it involves processing and selecting the collected attributes and information optimally. In this modeling process, we utilize the weighted arithmetic mean operator. For a given set of data, the following are the attributes:

$$(a_1, a_2, a_3, \ldots, a_n), a_i \in \mathbb{R} \tag{3}$$

If there is $WAA_W(a_1, a_2, a_3, \ldots, a_n) = \sum_{i=1}^{n} w_i a_i, w_i \in [0, 1]$, where $(w_1, w_2, w_3, \ldots, w_n)$ is the weight vector of attribute $(a_1, a_2, a_3, \ldots, a_n)$, then WAA_W is called the weighted arithmetic mean operator. In the Multi-attribute Decision Making model, there are various types of variables, including utility-type, cost-type, fixed-type, deviating-type, interval-type, and deviating interval-type variables. The attributes of each resource are determined based on the requirements of computing scenarios. Considering the limited availability of computing resources in practical experimental environments, the value range of resource quantities is also determined based on the requirements of computing scenarios. The next step is to determine the weights for each attribute based on the specific scenario. In this article, we will define the weights for different business scenarios. They are listed in Table 1.

2.2 Computing Power Identification and Tracking

After converting user requirements into computing tasks, in order to achieve end-to-end tracking and ease of deployment, this paper establishes identification mappings at three levels of the system. In the service layer, computing service identification (SID) and computing service description identification (SBD) are used to describe the service type and task details for user interactions. In the resource adaptation layer, the computing cluster identification (FID) and computing cluster service description identification are utilized to identify the type of computing cluster and provide detailed information about the computing resources. In the network component layer, network component identification (NID) and network component description identification (NBD) are employed to identify the network components and provide specific information about them. Ultimately, modeling is performed for individual user requirements and can be represented as $S_M =< SID, SBD, FID, FBD, NID, NBD >$.

2.3 Computing Power Orchestration and Management

Computing power service chain orchestration involves segmenting computing tasks, combining computing resources and network topology, and considering the interests of various parties in the network to efficiently utilize computing resources and maximize the utility of the computing power service chain. This paper makes a breakthrough in the modeling of comprehensive consideration of multiple types of resources and constructs a multi-dimensional heterogeneous computing orchestration and management model.

Table 1. Weights of heterogeneous computing power in different scenarios.

Computing Power Scenario	Computing power requirements	Isomeric computing power	Weight
training scenario	PFLOPS-level computing power, More than 1 TB of storage capacity	[FLOPS, space_ava]	[0.5,0.5]
Reasoning-type scenarios	TFLOPS-level computing power, Less than 60 ms delay, more than 250 MB storage capacity	[FLOPS, delay_eva, space_ava]	[0.5,0.45,0.05]
AR/VR scenarios	Less than 50 ms delay, more than 4G memory, more than 32G storage capacity, EFLOPS-level computing power	[delay_ava, mem_ava, space_ava, FLOPS]	[0.35,0.15,0.2,0.3]
Video-like scenarios	GFLOPS computing power, less than 50 ms delay, more than 20 Mbps bandwidth and more than 16G memory	[FLOPS, delay_eva, eth_speed,mem_ava]	[0.2,0.3,0.1,0.4]
Intelligent driving scenarios	TFLOPS-level computing power, 5-10 ms-level delay, more than 16 gigabytes of memory, more than 128 gigabytes of storage capacity	[FLOPS, delay_eva,mem_ava,spaca_ava]	[0.4,0.4,0.1,0.1]

Table 2 lists the key symbols used in the system modeling. Among them, N represents the set of computing power users, $N = \{1, 2, 3....., n\}$ represent the presence of n computing power users in the system. S is the set of computing nodes. $S = \{1, 2, 3....., s\}$ represent the presence of s computing nodes in the system. $\alpha = \{\vec{\alpha_1}, \vec{\alpha_2}, \vec{\alpha_3}, ..., \vec{\alpha_n}\}$ represents the overall offloading policy matrix for all users. $\vec{\alpha_k} = \{x_{k1}, x_{k2}, ..., x_{ks}\}$ represents the offloading policy vector for user k, which is one column of matrix α. The sum of $\{x_{k1}, x_{k2}, ..., x_{ks}\}$ being equal to 1 means that it represents the proportion of tasks offloaded to each computing node. The uplink data rate is represented by $R = B_u/n$, where B_u is the network port speed of the computing node, and n is the number of users accessing requests under the computing node. $T_k = \{R_k, C_k\}$ represents the computational task of user k. P_s represents the pricing strategy of computing node s. Detailed information about the symbols is provided in Table 2.

Table 2. Key symbols of the system model

variable	describe
N	user set
S	Calculate Node Set
α	The user unloads the policy matrix
$\vec{\alpha_k}$	Unload policy vector for user K
R	Data uplink rate
B_u	Calculate the network port rate of node u
T_k	Computational tasks for user k
C_k	Number of cpu revolutions required to process user k tasks
R_k	Task size for user k
t_k	Total latency to complete user k tasks
f_s	Compute the average cpu revolutions for node s
$Score_s$	Computational power evaluation of node s
P	Pricing Strategies for Computing Power Nodes
M	Computing Power Scenario

In a computing service, the total latency for service user k can be represented as the sum of upload latency, computation latency, and transmission latency. Since the transmission latency of the computation result can be negligible compared to the size of the computation task, this paper does not consider the transmission latency of the computation result. Therefore, the total latency for user k can be expressed as

$$t_k = t_{up} + t_c \quad (4)$$

where t_{up} represents the upload latency of the user's task and t_c represents the computation latency of the user's task. The latency of task distribution among computing nodes can be considered as a part of the task's upload latency. Let a represent the sum of latencies for processing subtasks at all nodes. Formula 4 can be further written as formula 5.

$$t_k = \frac{R_k}{R} + \sum_S \frac{x_k C_k}{f_s} \tag{5}$$

When users utilize computing power, they incur certain costs. In scenario M, the profit function of computing node s is denoted as $C_{M,s}(t)$, expressed as formula 6.

$$C_{M,s}(t) = (P_s - E_s) \cdot u_s \tag{6}$$

The pricing strategy (P_s) of a computing node in the expression can be represented as $\beta \cdot Score_s$, where β is a constant and $Score_s$ represents the computing power evaluation of computing node s. E_s is the pricing strategy of computing node s, which influences the user's willingness to use it. u_s represents the time consumed by computing node s in processing tasks. It is evident that higher pricing of a computing node leads to lower user willingness, and $E_s = (1/e)e^{P_s}$ represents the impact of pricing strategy on offloading willingness. Furthermore, u_s can be further decomposed into the sum of the time taken by computing node s to process all subtasks from each user on that node. Formula 6 can be further expressed as formula 7.

$$C_{M,s}(t) = (\beta \cdot Score_s(t) - \frac{1}{e}e^{P_s}) \cdot \sum_K \frac{x_{ks} C_k}{f_s} \tag{7}$$

Considering the user's profit, from the user's perspective, they aim to fulfill the computational needs of a task in the shortest possible time and with minimal costs. Users may choose multiple computing nodes to achieve this goal. Let the cost function of the user in computing scenario M be denoted as $C_{M,k}(t)$, as shown in formula 7, where $C_{M,k}(t)$ can be represented as:

$$C_{M,k}(t) = \beta \cdot Score_s(t) \cdot \sum_s \frac{x_{ks} C_k}{f_s} = \sum_s \frac{P_s x_{ks} C_k}{f_s} \tag{8}$$

Considering the presence of computing users and computing nodes in the system, we define a computing power provider where users obtain computing power from to solve their tasks. The computing users are required to pay fees to the computing power provider, who sets the pricing strategies for each computing node. The computing power service chain orchestration module should take into account both the profit of the computing power provider and the benefits of the computing users. Based on this consideration, the system is modeled as a Stackelberg model.

Let there be N computing users and one computing power provider. The strategy adopted by the provider is to determine the pricing for each computing node, represented as the pricing set $P = \{P_1, P_2, ..., P_s\}$. The strategy adopted by the computing users is to allocate a collective task to multiple different computing nodes, with the allocation proportions denoted as $\vec{\alpha_k} = \{x_{k1}, x_{k2}, ..., x_{ks}\}$. For N users, their strategies can be combined into a matrix $\alpha = \{\vec{\alpha_1}, \vec{\alpha_2}, \vec{\alpha_3}, ..., \vec{\alpha_n}\}$. The objective of the computing power provider is to maximize its profit, while the objective of the computing users is to minimize the combined cost of latency and fees. The Stackelberg game for computing task allocation and computing node pricing in the computing power service chain orchestration scenario can be represented as $G_M = \{\alpha, P, V(\alpha), V(P)\}$. Here, $V(\alpha)$ and $V(P)$ represent the utility functions of the computing users and the computing power provider, respectively. $V(\alpha) = \{V_1(\vec{\alpha}_1), V_2(\vec{\alpha}_2), ..., V_n(\vec{\alpha}_n)\}$ represents the set of utility functions for the N computing users.

The utility function of user k is related to the resources required to complete the task and can be represented by formula 9.

$$V_k(\vec{\alpha}_k) = \gamma \log \prod_s (2 + x_{ks}) - i_t t_k - C_{M,K}(t) \tag{9}$$

$\gamma \log \prod_s (2 + x_{ks})$ represents the utility function of the user under the offloading strategy $\vec{\alpha}_k$, and $\gamma = ce^{P_s/c}$, $c > 0$ is a function related to the user experience. $i_t, i_t \in (0, 1]$ is a latency-sensitive constant. Combined with the aforementioned time cost, the expansion can be completed.

$$\begin{aligned}V_k(\vec{\alpha_k}) &= ce^{\frac{P_s}{c}} \cdot \log \prod_s (2 + x_{ks}) - i_\tau(\frac{R_k}{R} + \sum_s \frac{x_{ks}C_k}{f_s}) - \sum_s \frac{P_s X_{ks} C_k}{f_s} \\ &= ce^{\frac{r_s}{c}} \cdot \log \prod_s (2 + x_{ks}) - \left(i_t \frac{R_k}{R} + i_t C_k \sum_s \frac{x_{ks}}{f_s} + C_k \sum_s \frac{P_s X_{ks}}{f_s}\right)\end{aligned} \tag{10}$$

Similarly, for the computing power provider, its utility function can be represented by formula 11.

$$V(P) = \sum_s (C_{M,s}(t)) \tag{11}$$

$V(P)$ represents the summation of the utility functions across all computing nodes. Expanding formula 11, we have:

$$\begin{aligned}V(P) &= \sum_s \left((P_s - \frac{1}{e}e^{P_s}) \sum_k \frac{X_{ks}C_k}{f_s}\right) \\ &= \sum_s (\frac{P_s - \frac{1}{e}e^{P_s}}{f_s}) \sum_k X_{ks}C_k\end{aligned} \tag{12}$$

In the Stackelberg model GM established for the computing power service chain orchestration scenario, the computing users can always find their corresponding Nash equilibrium. Therefore, the algorithmic objective for solving the Stackelberg game model $G_M = \{\alpha, P, V(\alpha), V(P)\}$ in the computing power service chain task orchestration scenario is to maximize the utility functions of both the computing power provider and the computing users. The goal of the algorithm is to solve this optimization problem.

$$\begin{cases} \max V_\mathbf{k}(\overrightarrow{a_\mathbf{k}})_{k \in N} \\ \max V(\mathbf{P}) \end{cases}$$
$$\text{s.t.} \quad 0 \leq X_{k,s} \leq 1, \forall X_{k,s} \in \overrightarrow{\alpha_k} \quad (13)$$
$$\sum_s X_{k,s} = 1$$

The architecture, numerous sub-functional modules, and design of the network's computing power service chain orchestration system are the key topics of this chapter. Overall, it follows the architectural principle of "computing power as a service" by incorporating it into the network. Through automated service deployment, modeling of computing capacity, and job allocation algorithms, the system, which makes use of programmable networking technologies, enables efficient exploitation of dispersed computing resources. Optimizing task latency and increasing resource efficiency are the objectives.

3 Algorithm Design

In this section, we introduce the main model methods that are utilized in the orchestration and management of computing power. They consist of the Nash equilibrium method for orchestrating the computing power service chain in the Stackelberg scenario and the optimal job allocation algorithm for computing power consumers. We refer to these algorithms collectively as CSFC-stackelberg (computing service function chaining stackelberg). These algorithms seek to determine the optimum computing job allocation strategy for each user under the initial pricing strategy as well as the best offloading method for each user.

The computing power service chain orchestration algorithm is solved using gradient iteration. T The precise method entails first identifying the best offloading plan for each user under a specified pricing scheme, after which the best pricing plan for computing is repeatedly solved. This is demonstrated in Algorithm 1 and Algorithm 2.

Algorithm 1 Optimal Task Allocation Algorithm for Computing Power Users
---
Input: The utility function $V_k(\alpha_k)$ for user k, Initial calculation force Pricing P, Maximum number of iterations max_Iter, iteration threshold eps

Output: The best uninstallation strategy α_k for user k

1 : Initialize learning rate φ, Initialize unloading decision vector x_init

2 : do Establish constraints cons :

3 :   for x_i in x_init

4 :   1) $x_i \in (0,1]$

5 :   2) $\sum_i x_i = 1 = 1$

6 :   end for

7 : Gradient $grad\_V = \partial V_k(a_k) / \partial V_{ki}$ of $V_k(a_k)$

8 : Find the initial value f_init of $V_k(a_k)$ at x_init

9 : for i in max_iter

10 : $x\_new = x\_init + \varphi \cdot \partial V_k(a_k) / \partial V_{ki}$

11 :   Make x_New satisfies cons

12 :   Find the initial value f_new of $V_k(a_k)$ at x_new

13 :   Find the rate of change of a function delta =f_new - f_init

14 :   if delta<=eps

15 :   return x_new

16 : end for

The goal of Algorithm 1 is to determine which offloading method is best for each particular user. First, the method defines the restrictions for the output user offloading decision vector using the input parameters. For the user, it determines the utility function's gradient. It updates the offloading decision vector using the formula in line 10 and determines the function value at the new offloading decision vector within the bounds of the maximum number of repetitions. The best computing job allocation method for user k has been identified, and the algorithm returns if the difference between the new and old function values is less than the iteration threshold.

Algorithm 2 makes use of Algorithm 1 to determine the best way to distribute computing tasks among all users under the initial pricing plan. When the change in the computing provider's utility function is less than the specified iteration threshold, the pricing strategy is then updated. The ideal computing pricing strategy and the ideal computing work allocation approach can be found in Nash equilibrium.

Algorithm 2 Stackelberg's Nash Equilibrium Algorithm for Computational Service Chain Orchestration Scenarios

---

Input: Initial calculation force pricing P, Number of users k, Calculate the number of nodes S, Iteration threshold eps

Output: Pricing P^* under Nash Equilibrium, Algorithm task allocation strategy A^*

1 : Initialize learning rate φ

2 : Initialize the computing power task allocation strategy matrix A

3 : for k_i in k

4 : Call Algorithm 1 to find the optimal unloading strategy a

5 : Add a to A

6 : end for

7 : Gradient $grad\_P = \partial V(P)/\partial V_i$ of P

8 : Find the initial value p_init of P

9 : for i in max_iter

10 : do

11 : $p\_new = p\_init + \varphi \cdot \partial V_k(a_k)/\partial V_{ki}$

12 : Find the optimal arithmetic allocation strategy A_new under p_new

13 : Find the rate of change of the function delta =p_new - p_init

14 : if delta<=eps

15 : $P^* = p\_new, A^* = A\_new$

16 : end if

17 : return P^*, A^*

18 : end for

---

4 Experimental Environment

The detailed implementation of the aforementioned modules is provided in this chapter, and the system testing is finished. The system uses seven servers with x86 architecture, one of which is used exclusively for the system engine and Zabbix server functions, and the other six are each running Zabbix clients. Docker containers hold the Zabbix clients that have been installed. On virtual computers, the deployment module for computing power service chains is tested. Mininet is used to build the network topology, with BMV2 switches deployed in the system for processing computing tasks and Mininet Xterm terminals serving as task containers. Solid lines denote data channels, whereas dashed lines denote control channels. The topology of the experiment is shown in Fig. 2. Table 3 introduces the system hardware configuration.

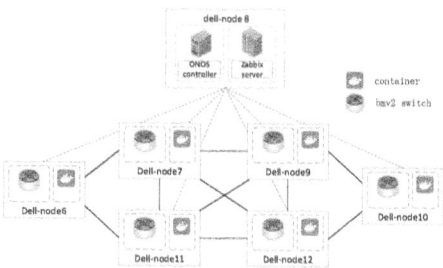

Fig. 2. Experimental Test Topology

Table 3. System Hardware Configuration

equipment	model
dell-node8	DELL EMC PowerEdge R740xd
Other dell-node	DELL EMC PowerEdge R640

5 Experimental Results

To verify the usefulness of these functional modules, functional testing is done in this part on the computing power awareness and modeling module, the identity tracking module, and the computing power service chain orchestration module in the network. For various user counts, the suggested CSFC-Stackelberg algorithm's convergence time is examined and tested.

Figure 3 shows the computational power provider's earnings for various user counts. It contrasts the average allocation strategy with the single-node execution strategy. The profit of the provider of computing power increases together with the growth of consumers. This is because the increase in the number of users leads to more computing power usage, which enhances the profit. However, the competition among users for limited computing resources may cause the computing power provider to raise prices. The CSFC-Stackelberg algorithm can maximize the computer power provider's profit.

User utility is the amount of computational work that a user can accomplish per unit of expenditure. Figure 4 contrasts the single-node execution method with the average allocation technique by comparing the average utility values of computing users. The

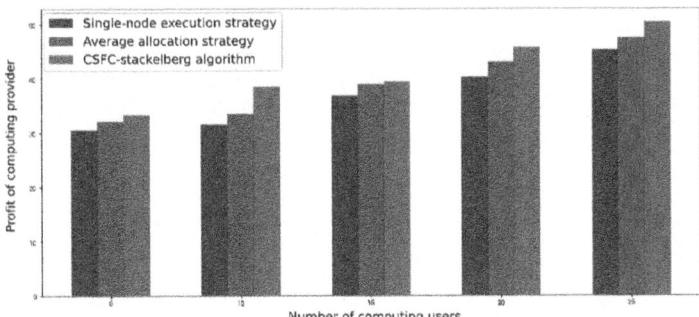

Fig. 3. Analysis of Benefits for Computing Power Providers.

CSFC-Stackelberg method, out of the three algorithms, optimizes the utility value for customers, as can be seen from the graph. This is due to the fact that the single-node computing strategy and average allocation approach do not account for the varied pricing of various computing nodes.

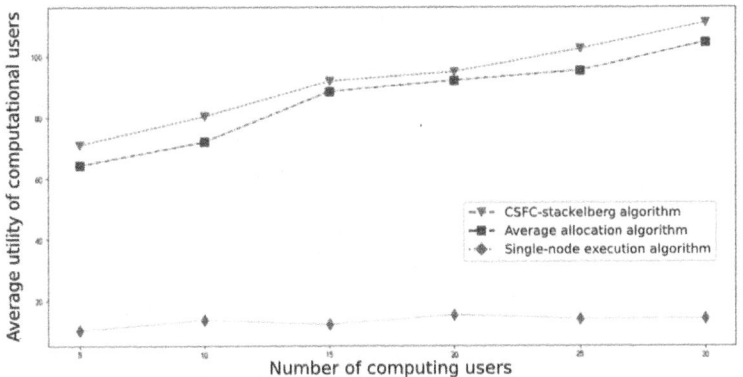

Fig. 4. Analysis of Benefits for Computing Power Users.

The relationship between the quantity of task offloading for computing users on the computing nodes and the number of algorithm iterations is compared in Fig. 5. There are 6 computing nodes and 5 computing users in the graph. The task offloading by users starts out randomly, but as the algorithm is orchestrated, it eventually stabilizes. The task offloading technique of users stabilizes after about 30 iterations. For various computing user counts, the algorithm's convergence time was evaluated. Overall, the algorithm worked well, converging for small user groups in less than a minute.

The modules of the computing service chain orchestration system were validated, and the results showed that all modules worked properly, showcasing the advantages of the algorithm proposed in this paper.

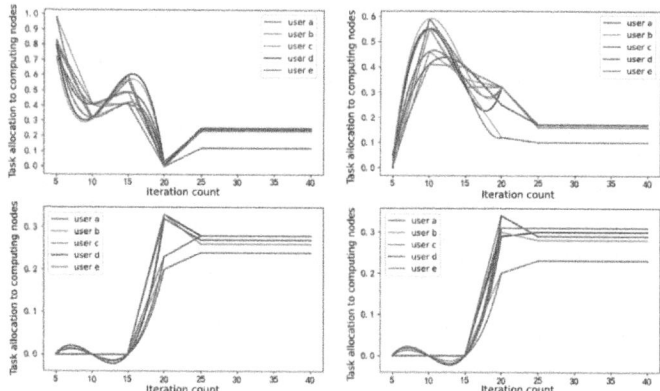

Fig. 5. Analysis of user offloading task volume

6 Summary

The computing power network technology is currently undergoing rapid development. This research investigated relevant technologies in the computing power network, including computing power perception modeling, identification management, and orchestration deployment. With a focus on efficiently handling computing tasks, we conducted in-depth research and proposed an implementable computing power service provider architecture model. This model enables multi-dimensional computing power measurement and modeling, establishing a standardized metric. Furthermore, we integrated computing power identification methods into existing network protocols, enabling effective tracking of users' computing tasks. By using the Stackelberg game algorithm, we were able to reduce user costs and achieve efficient processing of computing tasks, ensuring service quality and meeting user demands.

Funding. This work is supported by the National Key R&D Program of China 2022YFB2902400, The Major Key Project of PCI under grant no. PCL2022Y04, Fundamental Research Funds for the Central Universities under grant no. 2023JBGP003, Open Research Projects of Zhejiang Lab under grant no. 2022QA0AB03, Nature and Science Foundation of China under grant no. 92167204, 62072030, 61471029, 61972026, and 92267301.

References

1. Tang, X., et al.: Computing power network: the architecture of convergence of computing and networking towards 6G requirement. China Commun. **18**(2), 175–185 (2021). https://doi.org/10.23919/JCC.2021.02.011
2. Bosshart, P., Daly, D., Gibb, G., et al.: P4: programming protocol-independent packet processor. ACM SIGCOMM Comput. Commun. Rev. **44**(3), 87–95 (2014)
3. Cugini, F., Gunning, P., Paolucci, F., et al.: P4 in-band telemetry (INT) for latency-aware VNF in metro networks. In: Optical Fiber Communication Conference. Optica Publishing Group, M3Z. 6 (2019)

4. Osiński, T., Tarasiuk, H., Rajewski, L., et al.: DPPx: a P4-based Data Plane Programmability and Exposure framework to enhance NFV services. In: 2019 IEEE Conference on Network Softwarization (NetSoft), pp. 296–300. IEEE (2019)
5. Chen, X., Zhang, D., Wang, X., et al.: P4SC: towards high-performance service function chain implementation on the P4-capable device. In: 2019 IFIP/IEEE Symposium on Integrated Network and Service Management (IM), pp. 1–9. IEEE (2019)
6. Chen, J., Cheng, X., Chen, J., et al.: A lightweight SFC embedding framework in SDN/NFV-enabled wireless network based on reinforcement learning. IEEE Syst. J. **16**(3), 3817–3828 (2021)
7. Song, F., Ma, Y., You, I., Zhang, H.: Smart collaborative evolvement for virtual group creation in customized industrial IoT. IEEE Trans. Netw. Sci. Eng. (to appear)
8. Song, F., Ma, Y., Yuan, Z., You, I., Pau, G., Zhang, H.: Exploring reliable decentralized networks with smart collaborative theory. IEEE Commun. Mag. (to appear)
9. Jero, S., Koch, W., Skowyra, R., et al.: Identifier binding attacks and defenses in software-defined networks. In: USENIX Security Symposium, pp. 415–432 (2017)
10. Liu, S., Chen, J., Gao, S., et al.: Customizable service identification method and system design for smart integration identifier network. In: Quan, W. (eds.) Emerging Networking Architecture and Technologies: First International Conference, ICENAT 2022, Shenzhen, China, 15–17 November 2022, Proceedings, pp. 62–75. Springer, Singapore (2023). https://doi.org/10.1007/978-981-19-9697-9_6
11. Farhadi, V., et al.: Service placement and request scheduling for data-intensive applications in Edge Clouds. In: IEEE INFOCOM 2019 - IEEE Conference on Computer Communications, Paris, France, pp. 1279–1287 (2019). https://doi.org/10.1109/INFOCOM.2019.8737368
12. Pandey, S., Hong, J.W.-K., Yoo, J.-H.: Q-Learning based SFC deployment on Edge Computing Environment. In: 2020 21st Asia-Pacific Network Operations and Management Symposium (APNOMS), Daegu, Korea (South), pp. 220–226 (2020). https://doi.org/10.23919/APNOMS 50412.2020.9236981
13. Caballero, P., Banchs, A., De Veciana, G., et al.: Network slicing games: enabling customization in multi-tenant mobile networks. IEEE/ACM Trans. Networking **27**(2), 662–675 (2019)
14. Zhang, Y., Lu, Y., Yang, G., et al.: Multi-attribute decision making method for node importance metric in complex network. Appl. Sci. **12**(4), 1944 (2022)

Towards Smart Stream Scheduler for Multipath QUIC in Heterogeneous Networks

Anyi Li, Xiangbin Liang, Tianshu Wang, and Baokang Zhao[✉]

School of Computer, National University of Defense Technology,
Changsha 410073, China
{ayli,liangxiangbin13,wangtianshu21,bkzhao}@nudt.edu.cn

Abstract. In recent years, with the rapid development of internet communication technology and the continuous increase of terminal users, the amount of data transmission carried by the internet has increased rapidly, and users' requirements for service quality have also become higher. Multipath QUIC (MPQUIC) can utilize multiple network interfaces of terminal devices for data transmission, increasing bandwidth utilization, improving network fault tolerance and robustness. However, heuristic schedulers based on fixed parameters or models cannot adapt to dynamic heterogeneous network environments.

In this paper, we proposed a Smart Stream Scheduler (SmSS) that combines with deep reinforcement learning Double Deep Q-Network (DDQN). SmSS uses DDQN to capture the relationship between environmental states and stream scheduling decisions, so as to match the most suitable stream with idle paths based on the current environmental state, thereby improving transmission performance. The experimental results based on Mininet show that in dynamic heterogeneous network environments, compared to the default scheduler LRF of MPQUIC, SmSS can reduce transmission delay by up to 8.2%.

Keywords: MPQUIC · Multipath transmission · Data scheduling · Deep reinforcement learning · Heterogeneous network

1 Introduction

In today's Internet environment, network transmission quality and efficiency have always been issues of general concern. Quick UDP Internet Connection (QUIC) [1] is a UDP-based transport layer protocol proposed by Google. With advantages such as 0-RTT connection establishment, flexible congestion control algorithms, and multiplexing, QUIC aims to solve the problems of TCP handshake delay and head-of-line blocking (HoLB) [2], greatly improving transmission efficiency and having significant application value.

A. Li and X. Liang—These authors contributed equally to this work and should be considered co-first

However, QUIC cannot effectively utilize multiple network paths for data transmission. Currently, communication devices widely support multiple network interfaces, such as WiFi, wired networks, or 5G. Multipath transmission technology allows data to be transmitted simultaneously on multiple network paths, thereby improving network bandwidth utilization, enhancing transmission reliability and security. Existing multipath transmission protocols, such as Multipath TCP (MPTCP) [3], have demonstrated the potential of multipath transmission technology. Therefore, De Coninck et al [4] proposed the Multipath QUIC (MPQUIC) [5] based on the QUIC.

In multipath data transmission, reasonable schedulers are the key to achieving excellent performance. In heterogeneous network environments, different paths may have different bandwidths, delays, and packet loss rates. When the transmission of data packets on one path is blocked, data packets on other paths must wait for the transmission of data packets on that path to complete, resulting in the entire transmission process being blocked and causing HoLB problems, which poses challenges for the design of schedulers.

Traditional heuristic schedulers are usually based on certain fixed features or models to make decisions, so they can achieve better performance in specific scenarios, but they cannot adapt to dynamic changes in the network environment. In a real network environment, with changes in the number of user terminals, the levels of congestion, available bandwidth, and delay of each path will also dynamically change. The decisions made by heuristic schedulers at present may not be suitable for the greatly changed network environment, resulting in a decrease in transmission performance and a reduction in user experience.

Furthermore, in the context of the rapidly developing Internet era, different applications have presented diverse user demands. For example, batch data transmission requires high bandwidth, large-scale online games require reducing packet disordered length to smooth out jitter, while small file transmission requires low delay. However, existing heuristic schedulers based on certain fixed features or models are unable to meet diverse user demands.

In conclusion, in dynamically changing heterogeneous network environments, it is an urgent key technical issue to design a multipath smart stream scheduler based on MPQUIC, which can accurately perceive the dynamic changes in network environment and comprehensively consider the quality of service demands from various upper-layer applications.Compared with traditional heuristic schedulers, schedulers based on deep reinforcement learning are more suitable for handling decision-making problems in complex network environments.

In this paper, we proposed a Smart Stream Scheduler (SmSS) that combines deep reinforcement learning Double Deep Q-Network (DDQN) to solve the above problems. SmSS models scheduling problems as Markov Decision Processes (MDP). It designs environmental states based on stream information and path characteristics, and constructs reward functions based on factors such as throughput and stream completion time. By using DDQN, SmSS can capture the relationship between environmental states and stream scheduling decisions.

Thus, based on the current environmental state, it matches the most suitable stream to idle paths, thereby improving transmission performance.

The remaining parts of this paper are structured as follows. Section 2 introduces existing multipath schedulers. Section 3 introduces the framework design of SmSS. Section 4 introduces the algorithm design of SmSS. Section 5 is the experimental evaluation, and Sect. 6 is the conclusion of the paper.

2 Related Work

In recent years, numerous researchers have conducted extensive research on schedulers for multipath transmission protocols and have proposed various schedulers. The following paper introduces the relevant research work on multipath schedulers from the following three aspects: schedulers based on Round-Robin strategy, schedulers based on path characteristics, and stream-aware schedulers.

2.1 Schedulers Based on Round-Robin Strategy

The Round-Robin (RR) scheduler is based on a cyclic list and selects each available path sequentially for data transmission. RR is simple to implement, easy to deploy, and evenly distributes data packets to each path, avoiding excessive utilization of certain paths and resulting in load imbalance. However, RR cannot schedule based on the actual load of each path and performs poorly in heterogeneous network environments.

2.2 Schedulers Based on Path Characteristics

Schedulers based on path characteristics combine features such as round-trip time (RTT), bandwidth, delay, and packet loss rate, etc. Choi et al. proposed the Weighted Round-Robin (WRR) scheduler [6] based on RR, which assigns weights to paths based on path characteristics and allocates traffic based on the weight of each path, providing better flexibility, load balancing, and fault tolerance performance. Another common scheduler is the Lowest-RTT-First (LRF) scheduler [7], which is the default scheduler for various multipath transmission protocols such as MPTCP and MPQUIC. LRF measures the RTT of paths and selects the path with the smallest RTT for data transmission. However, LRF does not consider other path characteristics, may result in longer transmission time [8]. Lim et al. proposed the Earliest Completion First (ECF) scheduler [9], which evaluates path availability based on factors such as RTT, bandwidth, and connection-level send buffer size of the path. Unlike LRF, when the sending window of the optimal path is zero, ECF can choose not to send on the suboptimal paths and instead wait for the optimal path to become available. By maintaining an estimated completion time for each path, ECF will allocate the data to the path with the earliest expected completion time for transmission. When transmitting a small amount of data, ECF can effectively utilize all available paths and reduce transmission time. However, when transmitting a large amount of data, it can only select suitable transmission paths to handle the remaining data when the transmission is close to completion, and there is no difference from LRF in other transmission processes, as multiple paths are in full operation mode.

2.3 Stream-Aware Schedulers

Building on ECF, Rabitsch et al. proposed the Stream-Aware Earliest Completion First (SA-ECF) scheduler [10]. By introducing the priority feature of HTTP2.0 [11] and assigning weights to each stream, dynamic scheduling can be conducted based on the weights of streams. Unlike ECF, SA-ECF first finds the two paths with the smallest RTT and selects the path with shorter transmission time by comparing the transmission times of data packets on the two paths. However, SA-ECF estimates the completion time of a stream based on the remaining size of data on the stream, rather than the overall size of the data. This leads to different streams greedily choosing the fastest transmission path, causing load imbalance in the entire network [12]. X. Shi et al. proposed the Flexible Stream (FStream) scheduler [13], which schedules streams for transmission on a single path, effectively avoiding packet reordering issues caused by multipath transmission in heterogeneous network environments and reducing the transmission time of critical streams. However, this approach sacrifices the bandwidth aggregation advantage of multipath transmission.

3 Framework Design

This section describes the framework design of the SmSS based on deep reinforcement learning DDQN. SmSS models scheduling problems as Markov Decision Processes (MDP). In MDP, the next state only depends on the current state and the action taken in that state, independent of the past states. Therefore, the Q-Learning algorithm [14] can be used to maintain a Q-table of size $s_t * a_t$ to store the rewards of taking action a_t in state s_t. Q-Learning updates the Q-table iteratively until it converges, then the Q-table can be used to guide the Agent in selecting the action with the highest expected reward in each state.

The advantage of the Q-Learning algorithm is that when the state and action spaces are low-dimensional and discrete, Q-values can be simply stored in a Q-table. However, when the state and action spaces are high-dimensional and continuous, it becomes extremely difficult to construct a Q-table with a large storage space, and the time spent searching within it will also increase significantly.

During the process of multipath scheduling, considering the complex high-dimensional continuous environment state and action space, SmSS replaces the Q-table in the Q-Learning algorithm with a neural network. We select the Double Deep Q-Network (DDQN) [15], which is improved based on the Deep Q-Network (DQN) [16], to solve the MPQUIC scheduling problem. SmSS utilizes a large amount of experience in the Replay Buffer to train the neural network, gradually learning the relationship between various environment states and scheduling decisions, and can therefore adopt suitable scheduling strategies in various network environments. Fig 1 shows the architecture of SmSS, which is mainly composed of three parts: Agent, State Collectors, and Replay Buffer.

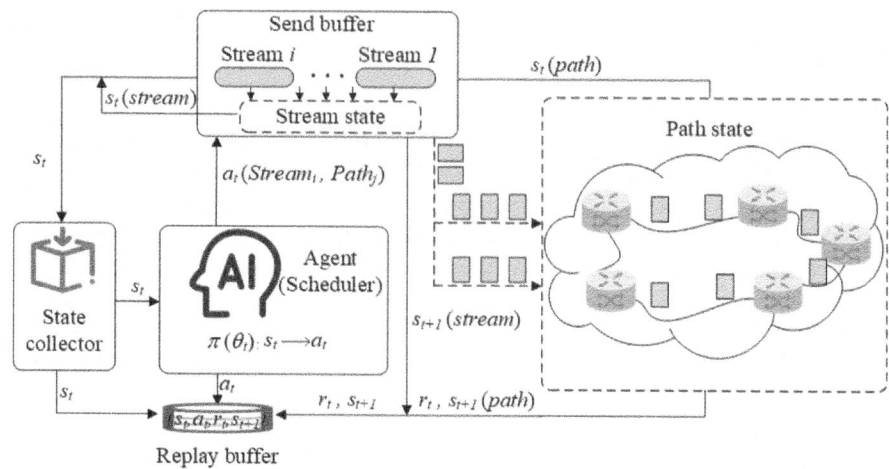

Fig. 1. Schematic diagram of the SMSS framework.

During the process of scheduling decision-making, the State Collector collects the state information $s_t(stream)$ of the stream and the path information $s_t(path)$ in the environment as the input of the Agent. Based on the input state information s_t, the Agent uses a neural network to output the most suitable action a_t, which is to match the appropriate stream to an idle path. After executing action a_t, the environment transitions to the next state s_{t+1} and generates a reward r_t. When the SmSS completes a complete scheduling process, it stores the acquired experience in the Replay Buffer in the form of (s_t, a_t, r_t, s_{t+1}) tuples. During the training process of the neural network, the SmSS samples batches of experience tuples from the Replay Buffer for training the neural network parameters. The specific training and scheduling process will be described in Sect. 4.

4 Algorithm Design

In this section, we employ a method based on deep reinforcement learning DDQN to address the data scheduling problem in MPQUIC. First, we map the stream scheduling task in MPQUIC to reinforcement learning in order to define its basic elements, then proceed to describe the update process of the DDQN algorithm and the subsequent stream scheduling process.

4.1 DDQN Stream Scheduling Model Definition

This section considers a typical scenario of multipath transmission, where the client uses the MPQUIC to request resources from the server. In this case, the server can divide the process of transmitting data through multiple paths into two steps. The first step is to schedule a portion of the stream data from the total buffer of the MPQUIC connection layer to the buffers of each path. The second step is to send the stream data from the buffers of each path to the receiver.

Two typical heuristic schedulers have been proposed in literature [17] and [18], based on the amount of data allocated to each path during each scheduling. The former adopts a package-based scheduling strategy, while the latter is based solely on the path parameters during scheduling, allocating concurrent streams in the total buffer all at once to each path. Therefore, the latter will generate the problem of over-allocation [19] and perform poorly when network conditions change. The remaining heuristic schedulers are somewhere in between. When attempting to combine reinforcement learning with scheduling decisions for each package, we found that it would result in a significant CPU load and instead affect transmission efficiency. Therefore, in order to strike a balance between CPU load and the performance disadvantage caused by over-allocation, we adopt a strategy between the two schedulers, which means when there is a path with a sending window size greater than zero, data is scheduled to each path atomically using a single stream.

During the process of multipath scheduling, we consider the case where there is a path available, which means the sending window of the path is greater than zero as the scheduling node. At each scheduling node t, the Agent, based on the environment state s_t composed of path and stream states, outputs stream scheduling action a_t that matches the most appropriate stream for the current available paths. The environment state s_t also changes continuously at different scheduling nodes, and the s_t between different scheduling nodes has the Markov property. Therefore, the entire scheduling process of MPQUIC can be modeled as a Markov decision process, and decision-making can be made during scheduling using a reinforcement learning-based model. Next, we will employ reinforcement learning to model the stream scheduling problem, mainly explaining the three fundamental elements of reinforcement learning: State, Action, and Reward:

1) State Space Definition: State refers to the network environment status information observed by the State Collectors in the SmSS framework, including data stream status information and path status information. At scheduling node t, the status information can be represented as s_t:

$$s_t = (p_t^1, p_t^2, ..., p_t^i, s_t^1, s_t^2, ..., s_t^j) \tag{1}$$

Among them, p_t^i and s_t^j respectively represent the status information of path i and stream j at scheduling node t. The path status information p_t^i includes multiple characteristics that depict the quality of the path and can be expressed as:

$$p_t^i = (cwnd_t^i, rtt_t^i, outgo_t^i, retrans_t^i) \tag{2}$$

In Eq. (2), $cwnd_t^i$, rtt_t^i, $outgo_t^i$, and $retrans_t^i$ respectively represent the congestion window, round-trip time, number of bytes sent but not yet acknowledged, and packet loss rate of path i at scheduling node t. The stream state information is defined as $s_t^j = size_t^j$, where $size_t^j$ denotes the size of stream j.

2) Action Space Definition: At scheduling node t, when inputting the network environment status information observed by the State Collectors, the Agent outputs Action a_t based on the scheduling strategy learned by the neural network. a_t represents the result of this scheduling, which selects the appropriate stream for transmission for the available path. As shown in Fig 1, a_t can be represented as a set of $(Stream_i, Path_j)$, meaning that stream i is scheduled to be transmitted on path j.

3) Reward Function Definition: After the Action output by Agent is executed, the network environment will enter the next State and provide feedback with a Reward r_t. This Reward is used to evaluate the quality of the previous stream scheduling Action. Therefore, the design of the Reward function should reflect the optimization objective of this paper. The Reward function is defined as shown in Eq. (3):

$$R(s_t, a_t) = \alpha \cdot Bdw_t^j - \beta \cdot Str_t^i - \delta \cdot \hat{D}_t \qquad (3)$$

Among them, $0 < \alpha, \beta, \delta < 1$, represent the weights of each reward factor. The detailed explanations of each reward factor are as follows:

Bdw_t^j represents the throughput of path j, which reflects one of the optimization goals of this paper. This factor allows paths with higher bandwidth to be fully utilized, effectively improving the transmission performance of MPQUIC.

Str_t^i factor is proportional to the completion time of stream i, so minimizing this factor can speed up the completion time of individual streams.

$\hat{D}_t$ is defined by Eq. (4):

$$\hat{D}_t = \frac{\sum_{i=1}^{k} d_i}{k} \qquad (4)$$

Among them, d_i represents the completion time of stream i. Therefore, $\hat{D}_t$ represents the average completion time of completed streams in a burst transmission, which allows users to obtain more requested resources in a shorter time. In addition, the Reward of this paper are calculated by the server itself at scheduling nodes, so the completion time of streams is estimated by adding the one-way delay of the path where the stream is located to the current time (at this time, the path has idle sending windows but the stream has not been completely transmitted yet). The main reason for calculating the reward value at the server side is as follows: on the one hand, if the Reward value is feedback from the client, due to network delay, when the server receives the reward sent by the client, it has already made the next scheduling Action, so there may be other scheduling actions between a scheduling Action and its corresponding Reward, which weakens the relationship between Action and Reward; on the other hand, transmitting the Reward between the client and the server requires modifying ACK frames and adding additional control traffic.

The goal of the Agent in reinforcement learning is to maximize the expected cumulative Reward $Q(s_t, a_t)$, and it is defined as follows:

$$Q(s_t, a_t) = \mathbb{E}[\sum_{t=0}^{\infty} \gamma^t R(s_t, a_t)] \tag{5}$$

Among them, γ is the discount factor, which is used to attenuate the contribution of future rewards to the current State.

4.2 Update and Stream Scheduling Process of DDQN Algorithm

As mentioned in Sect. 3, when the State and Action spaces are discrete and low-dimensional, the Q-learning algorithm can use a Q-table to store Q-values, which is simple and effective. However, the MPQUIC scheduling has a high-dimensional continuous State and Action space. In this case, Deep Reinforcement Learning (DRL) can use a deep neural networks to fit the Q function, it fits the accurate Q function by continuously training the neural network, thereby avoiding the problem of "curse of dimensionality". In this paper, we choose to use the Double Deep Q-Network (DDQN) [15], improved based on the Deep Q-Network (DQN) [16], to solve the MPQUIC scheduling problem.

In DQN, a deep neural network is used to fit the Q function, referred to as the Q-Network. Based on the weights θ of the layers in the Q-Network, the Q function can be represented as $Q(s, a; \theta)$. However, in the process of using a deep neural network to fit the Q function, the State transitions are continuous, so the sampled data is usually strongly correlated, which affects the efficiency of neural network updates. In addition, the process of updating parameters when fitting the Q function with a single deep neural network can be unstable, resulting in algorithm instability. Therefore, DQN solves the above problems by introducing the Experience Replay (ER) mechanism and fixed target Q-Network.

As shown in Fig. 2, on one hand, DQN stores the samples (s_t, a_t, r_t, s_{t+1}) obtained by interacting with the environment in a Replay Buffer. During training, a batch of data is randomly sampled from Replay Buffer for updating the neural network parameters θ, which greatly reduces the correlation between training samples and improves the utilization of samples. On the other hand, DQN uses two neural networks with same structure but different parameters: the Q-Network and the target Q-Network. Specifically, the Q-Network outputs the Q function value $Q(s, a; \theta)$ for the current State and Action, where θ is the Q-Network parameters. While θ^- represents the target Q-network parameters, and the target Q-network can further output the target Q function value of y^{DQN}:

$$y^{DQN} = r + \gamma \max_{a_{t+1}} Q(s_{t+1}, a_{t+1}; \theta^-) \tag{6}$$

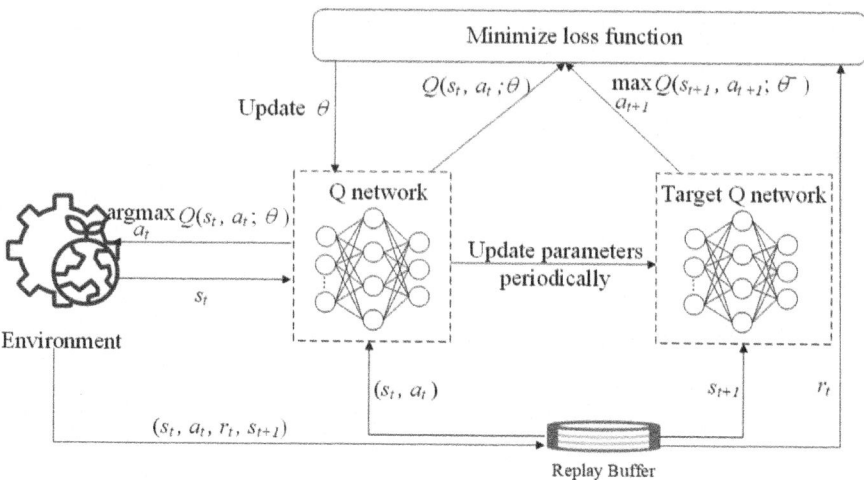

Fig. 2. DQN training process.

During the process of training, the Q-network is updated at each step. After a certain number of steps, the parameters of the Q-network are copied to the target Q-network, which ensures that the target Q-network remains stable for a certain period of time. This also reduces the correlation between the current Q-function values and the target Q-function values. Using the Q-network and the target Q-network, DQN updates the parameters of the Q-network by minimizing the loss function, in order to train the neural network. The loss function is defined as the expected mean square error between the target Q-function values and the Q-function values, and it can be represented by:

$$L(\theta) = \mathbb{E}[(r + \gamma \max_{a_{t+1}} Q(s_{t+1}, a_{t+1}; \theta^-) - Q(s_t, a_t; \theta))^2] \qquad (7)$$

However, there will be errors when using neural networks to predict Q_{max}. In DQN, when using the Eq. (6) to output the target Q function value, the neural network is also updated towards the target Q function value with the maximum error each time, leading to the problem of Overestimation. To solve this problem, DDQN selects the Action with the maximum Q function value using the Q-Network and evaluates the Q function value of that Action using the target Q-Network. Therefore, in DDQN, the Eq. (6) is replaced with the following form:

$$y^{DDQN} = r + \gamma Q(s_{t+1}, \arg\max_a Q(s_{t+1}, a; \theta); \theta^-) \qquad (8)$$

By substituting Eq. (8) into Eq. (7), the loss function in DDQN can be derived as Eq. (9):

$$L(\theta) = \mathbb{E}[(r + \gamma Q(s_{t+1}, \arg\max_a Q(s_{t+1}, a; \theta); \theta^-) - Q(s_t, a_t; \theta))^2] \qquad (9)$$

DDQN typically minimizes the loss function by using gradient descent to calculate the partial derivative of θ, in order to update the parameters θ of the Q-Network. The pseudocode for the training and scheduling process of the smart stream scheduler based on deep reinforcement learning DDQN proposed in this paper are as follows:

Algorithm 1: Training and scheduling process of smart stream scheduler based on DDQN.

Require: Discount factor: γ; Random action exploration rate: ϵ; Target Q-network update steps: C
Ensure: Stream scheduling policy
1: Initialize Replay Buffer D with capacity N
2: Initialize Q-network with random parameters θ
3: Initialize target Q-network with parameters θ of the Q-Network
4: **for** $episode = 1, M$ **do**
5:   Initialize State s_0 and preprocess to obtain feature input $\phi_0 = \phi(s_0)$
6:   **for** $t = 0, 1, ...T$ **do**
7:     With probability ϵ, select a random Action a_t
8:     Otherwise select Action $a_t = \underset{a}{argmax}\, Q(\phi(s_t), a; \theta)$
9:     Select appropriate stream based on Action a_t for available paths
10:    Obtain Reward r_t, preprocess next State s_{t+1} to obtain feature input $\phi = \phi(s_{t+1})$
11:    Store sample $(\phi_t, a_t, r_t, \phi_{t+1})$ in Replay Buffer D
12:    Sample a mini-batch of samples from Replay Buffer D
13:    $y_j = \begin{cases} r_j, & \text{if } episode \text{ ends at step } j+1 \\ r_j + \gamma Q(\phi_{t+1}, \underset{a}{argmax}\, Q(\phi_{t+1}, a; \theta); \theta^-), & \text{otherwise} \end{cases}$
14:    Update Q-network parameters θ by minimizing the loss function
15:    Update target Q-network parameters θ^- every C steps
16: **end for**
17: **end for**

Step 1: Initialize Replay Buffer, Q-Network, and target Q-Network parameters. Initialize the State based on the path and stream information, and preprocess to obtain the State feature vector.
Step 2: Output actions according to the State and $\epsilon - greedy$ mechanism in the Q-network. The $\epsilon - greedy$ avoids the decision of the Agent falling into local optimum by selecting random actions with a probability of ϵ. As the learning progresses, the neural network gradually stabilizes and ϵ decreases.
Step 3: Execute the selected Action, select the suitable stream for transmitting along the available paths.
Step 4: The Agent receives a Reward and the environment transitions to a new State. Store the previous State, the executed Action, the Reward, and the new State into Replay Buffer as a tuple.

Step 5: By using gradient descent, the Q-network parameters are updated by minimizing the loss function, and every certain number of steps, the parameters of the Q-network are copied to the target Q-network.

Repeat the training process until the neural network stabilizes, which indicates the completion of the training process. Afterwards, the trained neural network is used for scheduling decisions.

During the training process, the reinforcement learning algorithm learns from a large number of experience samples. Combining deep neural networks with reinforcement learning algorithms helps to avoid the "curse of dimensionality" problem in high-dimensional continuous state spaces. The reinforcement learning algorithm used in this paper is DDQN, which is an improvement of the DQN algorithm to address the issue of overestimation of Q function values. Finally, the $\epsilon - greedy$ mechanism continuously explores new possible scheduling actions to avoid getting stuck in local optima in scheduling decisions.

5 Experimental Analysis

5.1 Algorithm Implementation and Experimental Setup

In this section, we first provide a detailed description of the modules and interaction process involved in implementing the SmSS on top of the MPQUIC. Secondly, we introduce the experimental environment settings.

1) We introduce the DDQN learning module, which is responsible for storing experience data, used for training neural networks and making scheduling decisions. The DDQN learning module in this paper is implemented in Python and is based on the TensorFlow framework [20]. During the initialization process, we define parameters such as State, Action dimensions, Learning rate, etc., and build the Q-Network and target Q-Network. When receiving the status information from the Server-Client information transmission and collaborative module, we use the Q-Network and $\epsilon - greedy$ mechanism to generate actions, then save the current State, Action, Reward, and the next State to be transmitted as an experience tuple in the Replay Buffer. Finally, when the Replay Buffer contains enough experience data, we train the Q-Network using the experience data and synchronize the Q-Network parameters with the target Q-Network every certain number of steps.
2) We introduce the Server-Client information transmission and collaborative module, which is responsible for cross-language information exchange and interaction between the MPQUIC implemented in Golang and the learning module implemented in Python. This module is implemented using the UNIX Domain Socket (UDS) protocol, which allows data transmission without going through the network protocol stack and performing packet encapsulation, decapsulation, checksum calculation, etc. It is more efficient compared to the Socket protocol.The information transmission and collaboration module transmits environmental information, Rewards, and Action information.

3) We introduce the environment information collection and reward calculation module, which is responsible for collecting the current environmental State information and computing the rewards generated by the last scheduling when scheduling decisions are needed, and providing the above information to the information transmission and collaborative module.

Next, we will introduce the experimental environment and related parameter settings: The processor used in experiment training process is Inter(R)Xeon(R)CPU E5-2603 v4 @1.70GHz processor, the operating system is Ubuntu Linux 16.04. Using Mininet to simulate the experimental environment, the network topology is shown in Fig. 3. The Reward function parameters are $\alpha = 0.5$, $\beta = 0.25$, and $\delta = 0.25$. Following the approach in [18], we fix the parameters for path SP1 and vary the parameters for path SP2 (such as bandwidth, delay, etc.) to demonstrate the heterogeneity between different paths and evaluate the performance of different schedulers in heterogeneous networks. Apart from setting SP2's parameter as the variable parameter in later experiments, other parameters are set as default values shown in Table 1. Additionally, instead of fixing the path parameters as in [18], we further define the fluctuation range of network conditions as 2%, to reflect the impact of dynamic network environments on the performance of schedulers in multipath transmission.

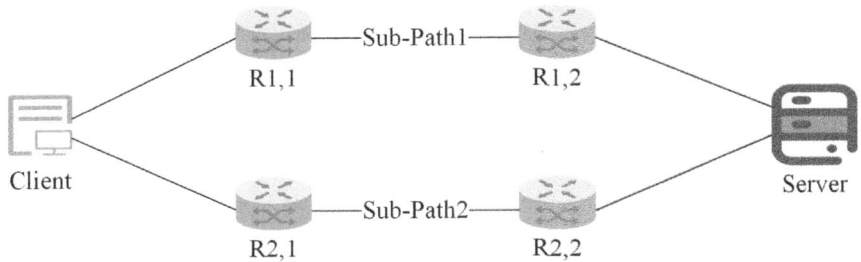

Fig. 3. A typical multipath network topology.

Several performance indicators were measured in our experiments in a dynamically changing heterogeneous network environment, including: 1) the completion time for all requested resources in the connection; 2) the average completion time for streams; 3) the length of disordered packets at the receiving end, which indicates the severity of HoLB for each scheduler. Severe HoLB can increase the completion time for requested resources and reduce user experience.

Table 1. Default experimental parameter settings.

Parameter	Value
SP1 Bandwidth	4 Mbps
SP1 End-to-End Delay	20 ms
SP1 Packet Loss Rate	0
SP2 Bandwidth	3 Mbps
SP2 End-to-End Delay	50 ms
SP2 Packet Loss Rate	1%
Network Condition Fluctuation Range	2%

5.2 Performance Evaluation

We evaluate the performance of SmSS by comparing it with the classical LRF and ECF. We first evaluates the performance of various schedulers in heterogeneous scenarios with low and high delay. The SP2 delay is randomly generated within the range of 40–60 ms and 80–150 ms in two different scenarios respectively.

Fig 4 shows the performance comparison results of the schedulers in heterogeneous scenarios with different delay. From Fig. 4, it can be seen that the median completion time of the SmSS is slightly smaller than the other two schedulers. This indicates that SmSS improves the overall performance compared to other schedulers in dynamic heterogeneous network environments with low delay.

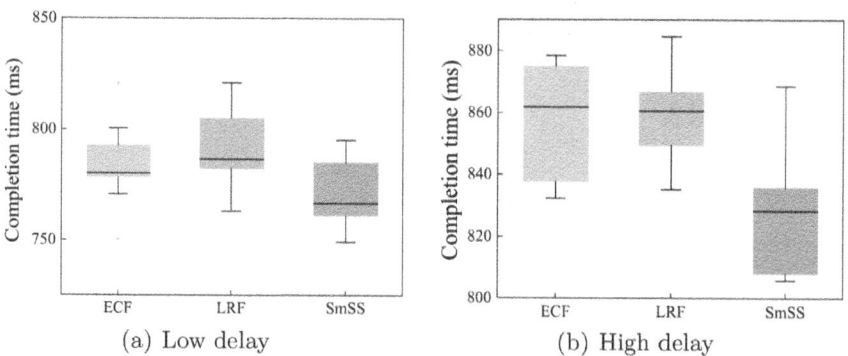

Fig. 4. The completion time of each scheduler in heterogeneous scenarios with different delay

As the level of delay increases, the completion time of all schedulers shows an increasing trend. In heterogeneous scenarios with high delay, compared to the other two schedulers, the performance improvement of SmSS is more pronounced. Because SmSS, based on a large amount of learned experience, can more fully consider the State and changes of the heterogeneous network environment with high delay, thereby scheduling the streams to the most appropriate paths for transmission. Meanwhile, as SmSS assigns each stream to a unique path, it effectively avoids HoLB caused by high path heterogeneity, which improves transmission efficiency.

Furthermore, we analyze the performance of various schedulers in heterogeneous scenarios with high and low bandwidth. SP2 bandwidth varies randomly from 0.5–1.5 Mbps and 2–4 Mbps in two different scenarios respectively.

Fig 5 shows the performance comparison results of the schedulers in heterogeneous scenarios with different bandwidth. It can be observed that compared to the completion time in heterogeneous scenarios with low bandwidth, the completion time for all schedulers significantly increases in heterogeneous scenarios with high bandwidth. This is mainly due to the fact that the bandwidth of SP2 is smaller in heterogeneous scenarios with high bandwidth, resulting in a total bandwidth smaller than the former. Moreover, in both scenarios, SmSS generally has a lower median completion time compared to the other schedulers. In the heterogeneous scenarios with high bandwidth, SmSS shows a more pronounced improvement in performance. Compared to the ECF and LRF, SmSS reduces the completion time by approximately 5.9% and 8.2% respectively, thus improving the transmission efficiency.

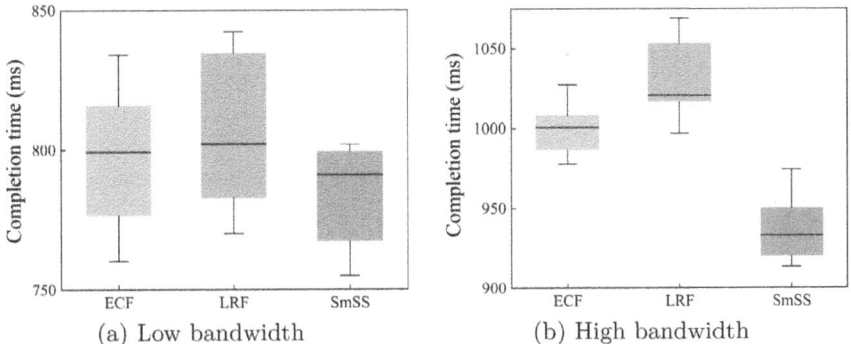

Fig. 5. Completion time of each scheduler in heterogeneous scenarios with different bandwidth

Speeding up the average completion time of streams means that users can access more requested objects in a unit of time, which can often improve user experience. Therefore, we use 10 burst streams as an example to test the performance of different schedulers in various scenarios. Since the test results in other heterogeneous scenarios with different bandwidth and delay are similar to those heterogeneous scenarios with low delay, we only presents the changes in the number of active streams over time and the corresponding average completion time of streams in heterogeneous scenarios with low delay, as shown in Fig. 6 and Fig. 7, respectively. From Fig. 6, it can be seen that SmSS quickly reduces the number of active streams. This is because the Reward function in SmSS aims to reduce the average completion time of streams, so it gives priority to scheduling smaller streams. While in the ECF and LRF scheduling processes, smaller streams may have to wait for the transmission of larger streams at certain moments, thus increasing their completion time. Furthermore, it is also evident from Fig. 7 that SmSS have significantly smaller overall average completion time of streams compared to ECF and LRF.

HoLB at the receiving end can cause jitter in the upper-layer applications [19], thus affecting user experience. Therefore, we tested the data disordered lengths generated at the receiving end by different schedulers in heterogeneous scenarios with different delay.

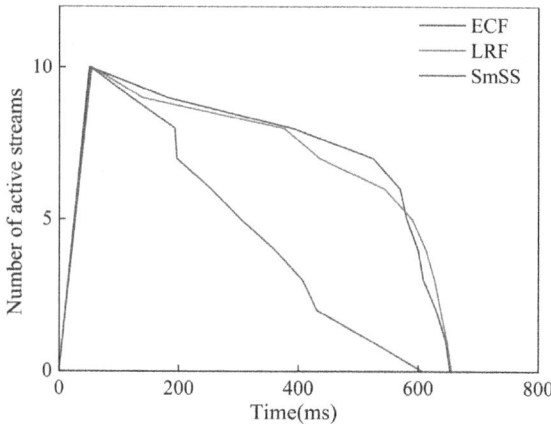

Fig. 6. Variation of the number of active streams in heterogeneous scenarios with low delay

For the purpose of comparison, Fig. 8 shows the disordered length of normalized data of different schedulers in heterogeneous scenarios with different delays. The disordered length of normalized data of a certain scheduler is defined as the ratio of the average data disordered length when using that algorithm to the average data disordered length when using LRF. From Fig. 8, it can be observed that as the level of delay increases, SmSS gradually show more advantages compared to LRF and ECF. This is because LRF and ECF schedulers do not consider the heterogeneity of path characteristics, and only schedule data in the selected path in order. Therefore, when the heterogeneity of multiple path characteristics increases, a large amount of disordered data will be generated at the receiving end, leading to more severe HoLB. In addition, compared to other algorithms, SmSS consistently maintains the lowest data disordered length in all scenarios. This is because SmSS assigns all data of a single stream to a unique path, effectively avoiding HoLB caused by heterogeneity in path characteristics. However, due to the dynamic changes in path conditions and packet loss, SmSS still cannot completely avoid data arriving out of order at the receiving end.

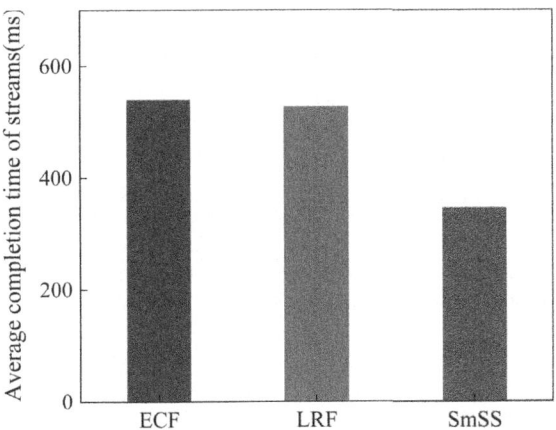

Fig. 7. Average Completion Time of streams in heterogeneous scenarios with low delay

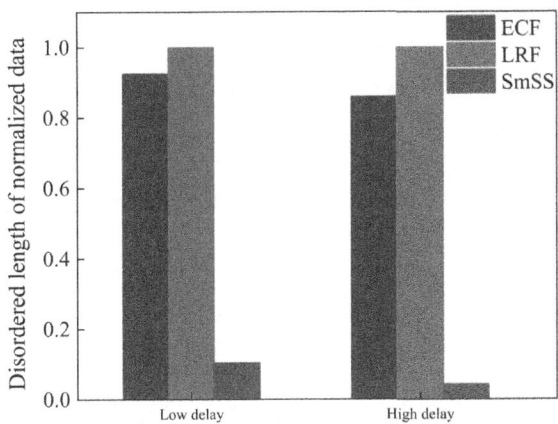

Fig. 8. Disordered length of normalized data in heterogeneous scenarios with different delays

6 Conclusion

In this paper, we proposed a Smart Stream Scheduler (SmSS), which combines deep reinforcement learning to address the poor performance of traditional heuristic schedulers in dynamic heterogeneous network environments. We first introduced the framework design of SmSS. Then presented the specific algorithm design, including modeling the stream scheduling task in MPQUIC as a Markov Decision Process, defining the elements in deep reinforcement learning, and the updating and stream scheduling process of DDQN algorithm. Finally, we evaluated the enhancement capability of SmSS on data transmission performance in dynamically heterogeneous network environments through experiments.

Acknowledgements. This work is supported by the National Natural Science Foundation of China under Grant No. U22B2005, 61972412.

References

1. Langley, A., et al.: The QUIC transport protocol: design and internet-scale deployment. In: ACM Special Interest Group on Data Communication, pp. 183–196 (2017)
2. Jonglez, B., Heusse, M., Gaujal, B.: SRPT-ECF: challenging Round-Robin for stream-aware multipath scheduling. In: 2020 IFIP Networking Conference, pp. 719–724 (2020)
3. Ford, A., Raiciu, C., Handley, M., Bonaventure, O.: TCP extensions for multipath operation with multiple addresses. RFC **6824**, 1–64 (2013)
4. De Coninck, Q., Bonaventure, O.: Multipath quic: design and evaluation[C]. In: international conference on emerging networking experiments and technologies, pp. 160–166 (2017)
5. Viernickel, T., Frommgen, A., Rizk, A., Koldehofe, B., Steinmetz, R.: Multipath QUIC: a deployable multipath transport protocol. In: IEEE International Conference on Communications 2018, pp. 1–7. Kansas City, MO, USA (2018)
6. Choi, K.W., Cho, Y.S., Lee, J.W., et al.: Optimal load balancing scheduler for MPTCP-based bandwidth aggregation in heterogeneous wireless environments. Comput. Commun. **112**(3), 116–130 (2017)
7. Paasch, C., et al.: Experimental evaluation of multipath TCP schedulers. In: 2014 ACM SIGCOMM Workshop on Capacity Sharing Workshop, pp. 27–32 (2014)
8. Shi, H., et al.: STMS: improving MPTCP throughput under heterogeneous networks. In: USENIX Annual Technical Conference 2018, pp. 719–730. Boston, MA, USA (2018)
9. Lim, Y.-S., et al.: ECF: an MPTCP path scheduler to manage heterogeneous paths [C]. In: 13th international conference on emerging networking experiments and technologies, pp. 147–159 (2017)
10. Rabitsch, A., Hurtig, P., Brunstrom, A.: A stream-aware multipath QUIC scheduler for heterogeneous paths. In: Workshop on the Evolution, Performance, and Interoperability of QUIC, pp. 29–35 (2018)
11. Wijnants, M., Marx, R., Quax, P., Lamotte, W.: HTTP/2 prioritization and its impact on web performance. In: World Wide Web Conference on World Wide Web 2018, pp. 1755–1764. Lyon, France (2018)
12. Shi, H., et al.: STMS: improving MPTCP throughput under heterogeneous networks [C]. In: USENIX Annual Technical Conference (USENIX ATC 18) 2018, pp. 719–730 (2018)
13. Shi, X., Wang, L., Zhang, F., et al.: Fstream: Flexible stream scheduling and prioritizing in multipath-quic [C]. In: IEEE 25th International Conference on Parallel and Distributed Systems (ICPADS) 2019, pp. 921–924 (2019)
14. Kumar, A., et al.: Conservative q-learning for offline reinforcement learning. Adv. Neural. Inf. Process. Syst. **33**(4), 1179–1191 (2020)
15. Van Hasselt, H., Guez, A., Silver, D.: Deep reinforcement learning with double q-learning. In: AAAI conference on artificial intelligence, pp. 2094–2100 (2016)
16. Mnih, V., et al.: Human-level control through deep reinforcement learning. Nature **518**, 529–533 (2015)
17. Hurtig, P., et al.: Low-latency scheduling in MPTCP. IEEE/ACM Trans. Networking **27**(1), 302–315 (2018)

18. Shi, X., et al.: PStream: priority-based stream scheduling for heterogeneous paths in multipath-QUIC. In: 29th International Conference on Computer Communications and Networks (ICCCN), pp. 1–8 (2020)
19. Liang, X., et al.: Towards Effective Multipath Scheduling with Multipath QUIC in Heterogeneous Paths. In: 10th International Conference on Information Systems and Computing Technology 2022, pp. 472–479 (2022)
20. Abadi, M.: TensorFlow: learning functions at scale. In: 21st ACM SIGPLAN International Conference on Functional Programming, pp. 1–1 (2016)

VotePipe: Efficient Heavy Hitter Detection in Programmable Data Plane

Danqi Li[1], Ningbo Tian[1], Kun Qiu[2], Harry Chang[2], Xiahui Yu[2], and Jin Zhao[1(✉)]

[1] School of Computer Science, Fudan University, Shanghai, China
{20307130353,jzhao}@fudan.edu.cn
[2] Intel Asia-Pacific Research & Development Ltd., Shanghai, China
{kun.qiu,harry.chang,xiahui.yu}@intel.com

Abstract. Heavy Hitter Detection (HHD) is of crucial importance in various applications, such as load balancing, traffic engineering, and DDoS attack detection. The emergence of programmable switches provides novel and effective solutions to heavy hitter detection by offloading network measurement tasks to the data plane. However, restrictions of the hardware architecture make it difficult for most HHD algorithms to be deployed on real devices. In this paper, we propose VotePipe, an HHD algorithm which completely deployed on the data plane of P4 programmable switches and circumvents the hardware restrictions. It adopts a flow count decay mechanism based on "flow age" to kick out outdated flows timely. Besides, it evicts small flows efficiently and minimizes replacement operations with a flow filtering and updating mechanism based on voting. We implement and deploy VotePipe on a real device and use 5 real Internet backbone network datasets and an Ixia traffic generator to test and compare the performance of VotePipe with 3 classic or state-of-the-art HHD algorithms that can be implemented on real machines. The results show that with acceptable memory consumption, VotePipe can achieve better accuracy and throughput than other algorithms, while running more stable and efficiently.

Keywords: heavy hitter detection · P4 · programmable switch

1 Introduction

The explosion of network traffic has boosted a great deal of research around network measurement, among which Heavy Hitters Detection(HHD) is of significant importance. Heavy hitters refer to a small number of flows that account for a majority of the network traffic. Identifying heavy hitters plays a crucial role in various applications, such as load balancing [13,21,25], traffic engineering [7,9], anomaly detection [16,17] and DDoS attack detection and mitigation [12,22]. Modern networks are characterized by their increasingly high bandwidth, with a rate of 10 Gbps to 100 Gbps for a single link. Additionally, traditional network devices have fixed control logic and forwarding logic, making it difficult to deploy and test new architectures and algorithms. The emergence of programmable switches addresses both issues and makes it possible to achieve improved network monitoring and management by offloading network measurement tasks to the data plane.

Traditional ways to detect heavy hitters include packet sampling [3,27] and polling technique [10]. Sampling-based methods normally use high sampling rates and polling-based methods adopt large polling intervals. Both techniques drop a lot of traffic characteristics and have limited accuracy due to the restricted storage memory and processing capacity of the CPU [26]. A later proposed data structure, sketch, strikes an ingenious balance of accuracy and resources [11, 18–20, 28]. Count-min Sketch [11] is a classic sketch-based scheme and is also a prototype of much relevant research. It applies d single-dimensional arrays to store the count value of network flows and uses d hash functions to hash the flow identifier of an incoming packet to get several values that correspond to positions in d counter arrays. The counters in these positions are incremented, and the minimum counting is reported as flow size.

However, sketch-based algorithms are difficult to implement on the data plane of programmable switches. For one thing, the value of d is greatly constrained by both memory size and memory access operations allowed per stage. Since it is impossible to parallelize calculations of dependent data, obtaining the minimum counter value requires multiple stages, which are limited in the programmable switches. For another, sketches cannot store flow identifiers, making it difficult for the control plane to proactively retrieve information regarding heavy hitters.

To address these challenges, we propose VotePipe, an HHD algorithm based on a hash table that entirely runs on the data plane of programmable switches. We first design a flow count decay mechanism that focuses on a single flow so as to reduce the count of flows that have not been accessed in the near term. We also present an efficient voting mechanism that attempts to filter out most small flows with two variables: positive votes and negative votes. Based on these two schemes, we implement the logic design of VotePipe and deploy it to hardware programmable switches. We evaluate VotePipe by comparing it against Count-min Sketch [11], PRECISION [4], and dSketch [15] in terms of accuracy, recall rate, false positive rate, false negative rate, throughput, memory overheads, and loopback packet ratio. Results show that VotePipe outperforms the other algorithms in all these metrics.

In summary, we make the following contributions:

- We design VotePipe, an efficient heavy hitter detection algorithm that can be fully implemented on the PISA-based programmable data plane without the intervention of the control plane, filling a gap in this field.
- We propose a flow count decay mechanism based on flow age and flow filtering and updating mechanism based on voting, which yields higher accuracy and recall rates while ensuring acceptable overhead costs.
- We successfully deploy VotePipe on a real hardware platform under the operating and memory access constraints of P4 programmable switches. The experimental results show that VotePipe can operate with high throughput and stability even under heavy traffic loads.

2 Background and Motivation

2.1 P4 Language and Programmable Switch

P4 [8], which refers to Programming Protocol-independent Packet Processors, is a language to describe the forward logic of packets in programmable data planes such as ASICs, NICs, and FPGAs [23]. The advantages of P4 include reconfigurability, protocol independence, and target independence [24].

Programmable switches are based on the P4 language and built using open networking hardware switches founded on off-the-shelf merchant silicon. They enable better performance in terms of power consumption and latency and can be programmed to introduce new features or perform various functions and tasks very quickly.

PISA (Protocol Independent Switch Architecture) is a packet processing model with a pipeline architecture, consisting of a programmable parser, a programmable match-action pipeline, and a programmable deparser [1]. The match-action pipeline contains multiple stages arranged in sequence, each with multiple memory blocks (tables, registers) and ALUs. The memory blocks and ALUs of different stages cannot access each other. It is allowed to simultaneously search and execute actions in a single stage, but dependent matches must be compiled into different stages. The PISA architecture is shown in Fig. 1.

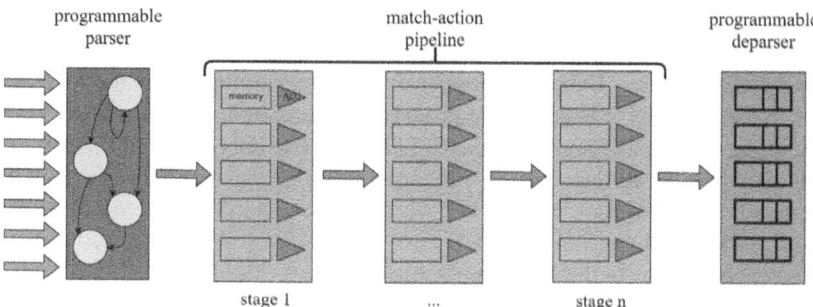

Fig. 1. PISA Architecture [1]

As shown in Fig. 1, pipeline architecture of the P4 language and PISA model leads to the following limitations on PISA-architecture-based switches [1]:

Limited Memory Access. As shown in Fig. 1, a fixed and limited amount of memory resources (TCAM and SRAM) and ALUs are allocated for each stage and cannot be accessed by other stages. The reason is that pipeline stages process different packets individually at one point, thus it may lead to read-write hazards if different stages are allowed to access the same memory unit simultaneously. Besides, a packet can only access each memory block(such as a table) once owing to the limitation of per-stage timing. To summarize, illegal accesses to memory resources across stages or multiple accesses to the same memory block are not allowed, which put forward higher requirements for programmers to satisfy.

Limited Actions and Branches in Each Stage. P4 does not support division, floating point operations, or loop operations. In each pipeline stage of PISA, only primitive arithmetic such as addition, subtraction, modulus, and shift operations can be executed. Each of these operations can only be executed once in a single stage. Additionally, branching operations within stages are limited (2 in Tofino [2] ASICs for instance) due to per-stage timing requirements.

Limited Processing Stages. To ensure Tbps-level throughput, programmable switches have a fixed and limited maximum amount of stages. For example, the maximum number of stages is 24 (12 for the input pipeline and 12 for the output pipeline) in Tofino ASICs. The programmer must not exceed the maximum limit when adding stages to conduct an algorithm. One solution to circumvent the restriction is to recirculate some packets, which has a bad impact on the throughput.

2.2 Related Work and Limitations

Heavy hitters are extensively found in DCN and ISP networks [6]. It is vital to identify them to provide helpful information for other network measurement applications. Traditional solutions include sampling and polling [3,10,27], which have trouble accomplishing fine-grained detection tasks. Collecting and processing every packet in network traffic can tackle the issue but with enormous expenses. The emergence of data structures like sketch and hash tables makes it possible to strike a balance between accuracy and memory consumption.

Sketch-Based Algorithm. Sketch is a probability-based data structure that utilizes hashing techniques to summarize incoming traffic flows, sacrificing acceptable accuracy for a small-scale data structure. Existing work includes UnivMon [19], Cold Filter [33], Elastic Sketch [31], Heavykeeper [32], etc. However, they are not intended for P4 programmable data plane and require many modifications before being deployed on programmable switches. Elastic sketch [31] designs a structure consisting of a heavy part and a light part that is adaptive to current traffic characteristics. But it can only be realized on a software platform. The P4 version developed by the author undergoes significant modifications from the original design scheme. The light part no longer uses the Count-min sketch and even lost its original adaptability feature. Sequential Zeroing [29] introduces the concept of the sliding window into the P4 programmable data plane. It assigns the packets in the sliding window to k sketches and deploys each in a stage. It resets the counters based on the First-In-First-Out principle. Nonetheless, the scheme requires an additional sketch to enable a sliding window, leading to low memory utilization. Furthermore, the overall stage count it necessitates is excessively high, rendering it unfeasible on P4 programmable switches. dSketch [15] overcomes the low accuracy problem of interval-reset algorithms by introducing a time-decaying algorithm and a recirculation scheme to update counter values. However, it does not address the storage of flow identifiers and cannot support active querying from the control plane. Moreover, excessive looping packets can impact switch throughput. According to our tests using five real data sets in Sect. 8, the required looping packets can be as high as 2.7% to 4.3%, which is a considerable expense.

Hash-Table-Based Algorithm. Hash table records currently active top-k flows with relevant information such as flow identifier and counting in each table entry. Existing work includes Space-saving [20], CSS [5], HashPipe [28], PRECISION [4], etc. The challenge is to implement the algorithms within the constraints of hardware programmable switches while ensuring the flows stored in the table pertain to large flows. Hash-Pipe [28] is the first HHD algorithm designed for the P4 data plane. It always inserts the incoming packet in the first stage and replaces the flow with a smaller count value whenever a hash collision occurs in any subsequent stage beyond the first stage. However, frequently performing replacement operations severely impacts network throughput. Moreover, count information of the same flow may be dispersed across different stages, leading to underestimation of flow counting and failure to report large flows. Additionally, HashPipe cannot be deployed on hardware platforms because of the memory access limitation of pipeline architecture. PRECISION [4] makes several improvements to make HashPipe implementable on hardware platforms. It uses a probabilistic recirculation mechanism to loop back a portion of packets, which then carry the information obtained from the previous pipeline processing to update the hash table. However, the switch throughput is significantly impacted by the probabilistic recirculation method, resulting in high overhead for very modest accuracy increases.

3 Design Goals

In this paper, we aim to implement an algorithm that is not only available on PISA-based programmable switches but also has a favorable trade-off between accuracy, throughput, and memory overhead. To this end, we work on the following aspects:

Filtering and Updating the Flows in an Effective and Efficient Manner. The most important factor in accurately detecting top-k flows in real-time networks is to store as many large flows as possible in the hash table. Small flows dominate large flows in a network by a wide margin, leading to numerous hash collisions. The replacement policies of HashPipe and PRECISION both result in a large number of replacement operations, which generate a lot of recirculated packets and reduce throughput. With the aim to evict small flows and minimize replacement operations, we design a voting scheme in this paper.

Decay Flow Count Effectively. It is critical to promptly remove out-of-date flows and only keep counts of the most current flows. Previous studies typically measure network traffic within a time frame or a predetermined number of packets, which results in many large flows being missed [14]. Sliding window algorithms either do not work well with programmable hardware data planes or sacrifice too much performance. dSketch offers a count decay technique. However, it still relies on time windows, and the versatility is restricted by the duration of the time window. Our work attempts to overcome the constraint of time windows and to design a more efficient count decay mechanism.

Available for Active Querying from the Control Plane. Previous work such as Count-min Sketch and dSketch do not store flow identifiers. Such algorithms can only estimate the size of the flow when receiving a packet, and report it if the flow count exceeds a preset threshold. It is important to give network administrators the ability to actively query the data of large flows in order to meet practical objectives. Therefore, VotePipe utilizes a hash table as its data structure to store flow identifiers.

Acceptable Memory and Throughput Overheads. SRAM resources in the data plane of hardware programmable switches are limited and evenly allocated throughout each pipeline stage. Also, memory units in each stage have to share the SRAM resources. Therefore, great thought must go into how SRAM resources are used and allocated. Moreover, even while the packet recirculation mechanism gets around the restriction that memory units (like registers) can only be accessed once, it also introduces the issue that having too many looped packets will negatively impact the switch's throughput. In this paper, we aim to keep the size of overheads within reasonable bounds.

4 VotePipe Design

4.1 Flow Count Decay Mechanism

In order to reduce the count of flows that have not been accessed in the near term, we develop a flow count decay mechanism that focuses on a single flow based on the idea behind dSketch. Previous works have considered time windows from a network-wide perspective, but the lifespans of individual flows are unpredictable, and it is not feasible to process all flows within the same period. The viewpoint of this paper is that judgment should be made on a per-flow basis, without taking time windows into account, but rather consider whether the flow requires decay. The emphasis is on the identification of outdated flows. We introduce "age threshold" (age_t) and "flow age" (age) to denote how long a flow has not been accessed. Whenever an incoming packet matches the hash table, the flow's age is reset to 0. When a flow reaches the age of age_t, it is deemed to be too old and has to have its count value decayed by deducting the flow decay value *dVal*.

Figure 2 illustrates how the flow count decay mechanism works. In Fig. 2a, a packet belonging to flow A is successfully matched to the corresponding table entry. Since the *age* field in the entry is 0, it remains unchanged. The *val* field, which counts the number of packets of flow A, is incremented by 1. In Fig. 2b, a packet belonging to flow E collides with the hash value of flow B in the table entry. The *age* value in the table entry is increased to 8, but since $age > age_t$, no further action is taken. In Fig. 2c, a packet belonging to flow C is successfully matched. The *age* field is reset to 0 and the *val* field is incremented by 1. In Fig. 2d, a packet belonging to flow G is hashed to the table entry whose flow identifier is D. The matching fails, and the *age* field in this table entry reaches 30 after being increased by 1. At this point, *age* exceeds the preset threshold age_t, so *val* is decayed by *dVal*. If *val* $<dVal$, the *val* field is set to 0.

4.2 Flow Filtering and Updating Mechanism

During the heavy hitter detection process, the data plane replaces a table entry in two scenarios. The first scenario involves a small flow occupying the table entry. This happens when a packet of a small flow occupies an initially empty table entry, or when a flow that could have grown into a large flow replaces a prior flow that was stored in the

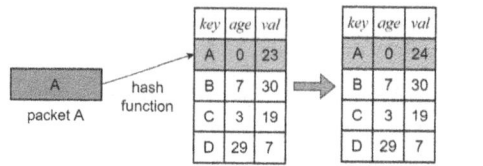
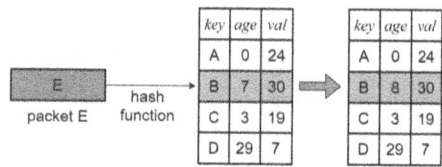

(a) Packet A matched and *age*=0. Increase *val* by 1.

(b) Packet E unmatched. Increase *age* by 1.

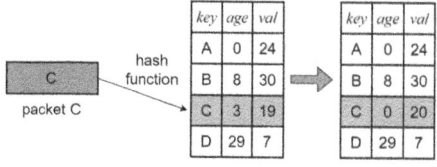
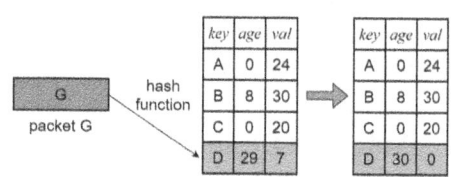

(c) Packet C matched. Reset *age* to 0 and increase *val* by 1.

(d) Packet G unmatched. Increase *age* by 1. Since *age* ≥ age_t, *val* decayed by *dVal*.

Fig. 2. Operations on *age* field ($age_t = 30$, $dVal = age$)

table entry but eventually fails to do so. The second scenario is when a large flow has ended. It must be thrown away in order to make room for a brand-new large flow or one that has the potential to grow into a large flow.

To accomplish goals 1 and 4, we believe that it is not necessary to store small flows that make up the majority of network traffic. If the flow in the table entry is replaced whenever a hash collision occurs, or with a certain probability, as in HashPipe and PRECISION, then there will be a massive number of replacement operations, resulting in significant additional costs. Therefore, our work employs the concept of cloning and recirculating packets to execute replacement operations while attempting to filter out the majority of small flows via a voting mechanism. Instead of using flow count value *val* described above, we use positive votes (v_y) and negative votes (v_n). When a packet arrives, a voting operation is performed based on whether its flow identifier matches the one that is currently stored in the table entry. If there is a match, a positive vote is added; otherwise, a negative vote is added. Once the partition of positive votes and negative votes reaches a preset value, the replacement operation will be triggered.

Figure 3 demonstrates the voting mechanism. It consists of two operations: the report operation and the replacement operation. When the positive vote of a table entry reaches the preset threshold *T*, it is deemed that the flow stored in this entry has developed into a large flow. The report operation is then initiated to send the flow identifier information and the count value of this flow to the control plane. On the other hand, when the negative vote of a table entry meets the replacement condition $v_n \geq \lambda v_y$ ($\lambda \in \mathbb{R}^+$), it is considered that the entry's flow is no longer large or cannot increase to become large. The replacement operation is then triggered, which updates the negative vote to 0 and replaces the corresponding value stored in the table entry with the flow identification and new positive vote (new_v_y) of the current packet.

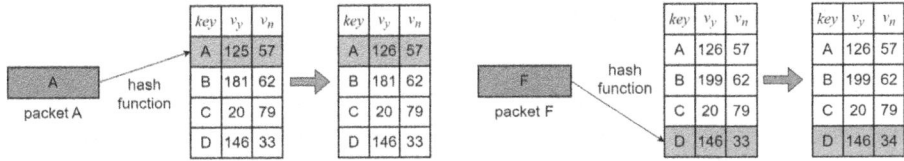

(a) Packet A matched, increase v_y by 1.

(b) Packet F unmatched, increase v_n by 1.

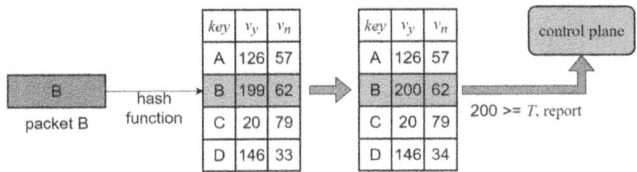

(c) Packet B matched, increase v_y by 1. Since $v_y=200 >T$, report B as a large flow.

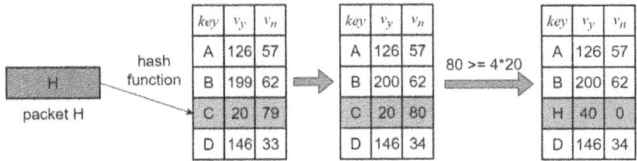

(d) Packet H unmatched, increase v_n by 1. Since $v_n \geq \lambda v_y$, replace the entry with (H,40,0).

Fig. 3. Voting mechanism ($T=200$, $\lambda = 4$, $new\_v_y = v_n \gg 1$).

As shown in Fig. 3a, a packet belonging to flow A is hashed to the first table entry. The matching is successful, and v_y of flow A is incremented by 1. In Fig. 3b, a packet belonging to flow F is hashed to the fourth table entry, which conflicts with the original flow identifier D. As a result of the hash collision, v_n is increased. In Fig. 3c, a packet belonging to flow B is hashed to the second table entry, where the matching is successful. At this point, v_y increases to 200, which satisfies the condition $v_y \geq T$. Consequently, the report operation is activated to report the flow identifier information and count information (v_y) of flow B to the control plane. In Fig. 3d, a packet belonging to flow H is hashed to the third table entry, where the matching fails. As a result, v_n increases to 80, which meets the replacement condition $v_n \geq 4v_y$. The replacement operation is then triggered to calculate the new positive vote $new\_v_y = 80 \gg 1 = 40$, as well as to replace the original value with the flow identifier H and the new positive vote 40, and the negative vote v_n is reset to 0.

5 VotePipe Architecture

5.1 Overall Architecture

This paper proposes a heavy hitter detection algorithm called VotePipe based on dSketch and PRECISION. The general design of VotePipe is shown in Fig. 4. VotePipe's detecting, updating, and reporting operations all take place in the data plane; the control plane only accesses the data plane when it has to perform a flow query. Whenever VotePipe detects a large flow, it reports the flow's details to the data plane. When an update operation is performed, VotePipe clones and loops back the packet to re-enter the pipeline for information update.

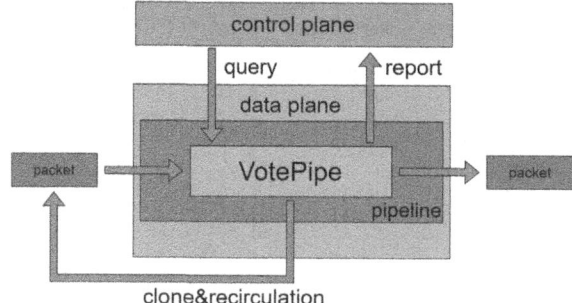

Fig. 4. VotePipe Architecture

index	*key*	*age*	v_y	v_n
0	key_1	age_1	v_{y1}	v_{n1}
1	key_2	age_2	v_{y2}	v_{n2}
2	key_3	age_3	v_{y3}	v_{n3}
...	...	...	...	...
k-1	key_k	age_k	v_{yk}	v_{nk}

Fig. 5. VotePipe Data Structure

5.2 Data Structure

VotePipe is a hash-table-based structure that combines a flow count decay mechanism based on "flow age" with a flow filtering and updating mechanism based on voting. The structure is represented by a single hash table with k entries, each of which has 5 fields: index (*index*), flow identifier (*key*), flow age (*age*), positive vote (v_y), and negative vote (v_n). This is depicted in Fig. 5. *index* represents the flow's position in the hash table. When a packet arrives, a hash function is employed to determine its corresponding index in the hash table. *key* uniquely identifies a flow. Its value can be extracted from the header fields of the packet. (In this paper, the packet's destination IP address serves as the flow identifier.) *age* represents the quantity of successive table entry flow misses. A greater flow age value implies that the flow has not been accessed for a longer duration. v_y represents the estimated count of packets for the flow stored in the table entry. It may decay as the flow age increases. v_n represents the total number of packets that have hash collisions with the flow stored in the current table entry. It not only effectively filters out small flows, but also provides initial v_y value for the new flow in the update operation.

5.3 Workflow

The pseudo-code of the VotePipe algorithm is shown in Algorithm 1, where H denotes the hash table used to hold the information for the flows.

VotePipe initially checks whether a packet is a looped packet after receiving it. If the answer is yes, the program extracts the data from the packet, updates the relevant table item with this data, and then discards the looping packet (lines 1–3). If not, VotePipe extracts the flow identifier of the packet and calculates its hash value to get the index of the table entry where the packet is hashed (line 5). If the table entry at *index* is empty, i.e., *key* equals 0, VotePipe directly puts the information of this packet into the table entry (lines 6–7). If the table entry at *index* is not empty, and the packet's flow identifier matches successfully, VotePipe then sets *age* to 0 and *matched* to True (lines 8–10).

Algorithm 1: Workflow of VotePipe in the Data Plane

Input: an incoming packet F

1. **if** *F.isRecirculatedPacket* **then**
2. $H[F.index].(key, age, v_y, v_n) \leftarrow (F.cKey, 0, F.cYes, 0)$;
3. $Drop(F)$;
4. **else**
5. $index \leftarrow H[F.key]$;
6.   **if** $H[index].key = 0$ **then**
7. $H[index].(key, age, v_y, v_n) \leftarrow (F.key, 0, 1, 0)$;
8.   **else if** $H[index].key = F.key$ **then**
9. $H[index].age \leftarrow 0$;
10. $matched \leftarrow True$;
11.   **else**
12. $H[index].age \leftarrow H[index].age + 1$;
13. $age \leftarrow H[index].age$;
14. $matched \leftarrow False$;
15.   **if** *matched* **then**
16. $H[index].v_y \leftarrow H[index].v_y + 1$;
17. $v_y \leftarrow H[index].v_y$;
18.   **if** $v_y \geq T$ **then**
19. $clone\_and\_report(F)$;
20.   **else**
21.   **if** $age \geq age_t$ **then**
22. $H[index].v_y \leftarrow \|H[index].v_y - dVal\|$;
23. $v_y \leftarrow H[index].v_y$;
24.   **else**
25. $v_y \leftarrow H[index].v_y$;
26. $H[index].v_n \leftarrow H[index].v_n + 1$;
27. $v_n \leftarrow H[index].v_n$;
28.   **if** $v_n \geq \lambda v_y$ **then**
29. $new\_v_y \leftarrow v_n >> 1$;
30. $F.(index, cKey, cYes) \leftarrow (index, F.key, new\_v_y)$;
31. $clone\_and\_recirculate(F)$;

Otherwise, VotePipe increases *age* by 1 and sets *matched* to False (lines 11–15). The *key* and *age* fields have now been processed in full.

The next step is to process the fields v_y and v_n. If *key* matches correctly, VotePipe increases the positive vote v_y by 1. If the modified v_y reaches the threshold T for large flow detection, the flow to which this packet belongs is considered a large flow. Then, VotePipe clones this packet, sends the cloned packet to the control plane for reporting, and continues to process the original packet (lines 16–21). If *key* does not match, VotePipe first checks if *age* has reached the age threshold age_t. If yes, a decay operation on v_y is performed (lines 23–25). If no, VotePipe obtains the current v_y value (lines 26–28). Next, VotePipe increases the negative vote v_n and checks if the updated value meets the replacement condition. If it does, VotePipe updates the corresponding entry inserting the necessary information into the packet's header field and cloning the packet. While the original packet continues with its usual processing, the cloned packet performs a loopback operation and re-enters the pipeline (lines 29–35). If the replacement condition is not met, no action is taken.

5.4 Control Plane

VotePipe completely manages network traffic in the data plane without the need for control plane involvement. The control plane is only responsible for receiving reported packets and requesting flow information. As a result, programmable switches can maximize their performance by hastening heavy hitter detection.

Algorithm 2: Workflow of VotePipe in the Control Plane

1 **Function** *Recieve(HH_table)*
2   **while** *True* **do**
3        $f \leftarrow getReport()$;
4        $HH\_table.add(f)$;
5   **end**
6 **Function** *Query(index)*
7     $f \leftarrow hash\_table.get(index)$;
8   **return** f;
9 **Function** *QueryAll(k, flow_table)*
10   **for** $index \leftarrow 0$ **to** $k-1$ **do**
11       $f \leftarrow hash\_table.get(index)$;
12       $flow\_table.add(f)$;
13   **end**

The control plane pseudo-code is provided in Algorithm 2, which offers three interfaces: *receive(HH_table)*, *query(index)* and *queryAll (k, flow_table)*. *receive ()* listens for heavy hitter's information reported by the data plane and stores it in *HH_table*. *query()* requests the details of the table entry at *index* in the data plane's hash table. *queryAll()* queries all of the information from the table entries in the data plane's hash table and stores it in *flow_table*.

5.5 Key Parameter Determination

This section mainly introduces how to determine the key parameters in VotePipe algorithm, including flow threshold T, flow age threshold age_t, flow count decay value $dVal$, flow replacement condition coefficient λ and new positive vote value new_val.

A programmable switch typically has two to four pipelines, and each pipeline is capable of deploying an algorithm. In this study, we investigate running the algorithm on a single pipeline. Suppose the maximum rate of the pipeline is v bps, the average link utilization rate is u, the average package size is s_a, the number of table entries is k, the fraction of the packet quantity to total packet quantity when a flow grows into the kth largest flow is τ, the hash function used to calculate the table index is uniform, then the value of T is as follows:

$$T = \left\lceil \tau \frac{uv}{8s_a k} \right\rceil$$

Under the premise that the hash function is uniform, the average number of packets that hit an entry per second is $\left\lceil \frac{uv}{8s_a k} \right\rceil$. If no packets pertaining to the flow stored in the entry arrive within t seconds, it is deemed that the flow age has to be decayed. Therefore, the value of age_t is

$$age_t = t \left\lceil \frac{uv}{8s_a k} \right\rceil$$

When the age of a flow reaches age_t, its count value needs to be decayed. dSketch [15] directly halves the count, resulting in the large flow being reduced prematurely to the point where it needs to be replaced, and the large flow may still be ongoing at this time. Therefore, instead of using half decay, we decrease the flow's count value by its age, i.e. $dVal = age$. This strategy has two advantages. First, when the flow stored in a table entry is no longer big, the decrease value increases as its age increases, so that it can meet the replacement condition faster. Second, if the flow stored within the table entry only temporarily blocks for a while, it can stop decreasing by refreshing the age field when its packet arrives.

In the algorithm, when the positive votes and negative votes ratio meets the condition $v_n \geq \lambda v_y$ (where λ is the replacement condition coefficient), the flow stored in the table entry needs to be updated. According to the analysis and experimental analysis of network traffic data sets by Yang T et al. [31], we recommend λ to take a value of 4, i.e. replacement condition is recommended for $v_n \geq 4v_y$.

In addition, when the replacement condition is met, given the composition of the count value in negative votes, it is recommended to take the half of v_n as the new positive vote's count value, i.e. $new\_val = v_n \gg 1$.

6 Implementation

6.1 Data Structure

In the PISA-based hardware programmable data plane, a stateful array is implemented via a set of SRAM-based registers. The SRAM resources are evenly distributed throughout each stage of the pipeline as shown in Fig. 6. Memory resources in different stages cannot access each other, and only one access to a position(*index*) of a register array can be made inside the same stage. The access is an atomic operation called *RegisterAction*. For a register array, up to four different *RegisterAction*s can be defined, but only one can be executed for each array during the lifespan of a packet. In each *apply()* function of a *RegisterAction*, two comparison ALUs and four arithmetic/logic ALUs are available. Consequently, register operations within a *RegisterAction* are severely constrained, making it difficult to design algorithms in hardware programmable data planes.

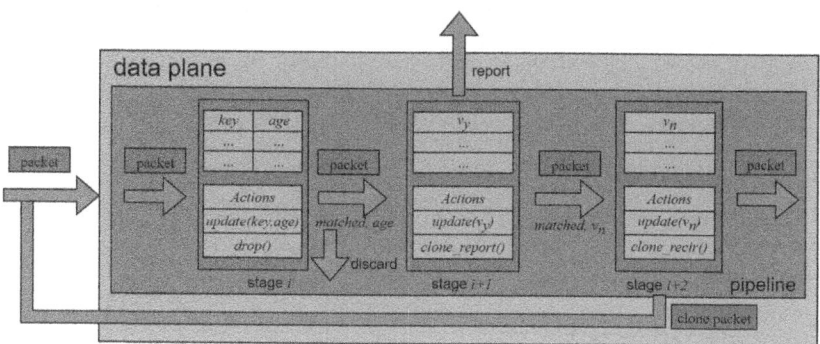

Fig. 6. Distribution of Memory Arrays and Associated Operations in VotePipe

As shown in Fig. 5, VotePipe has five fields in the hash table, namely *index*, *key*, *age*, v_y, and v_n. *index* is maintained by the data plane itself, while the other four are represented by register arrays. Registers support type *pair* <*value1*, *value2*>. Same as storing a single value, access, modification, and other operations on the two values still need to be completed in a single *RegisterAction*. *key* and *age* are stored as a pair in a register array. v_y and v_n cannot be placed in the same pair according to Algorithm 1, thus being implemented using two separate register arrays. To sum up, the data structure of VotePipe is implemented by three register arrays, all with k entries, representing the maximum number of flows that can be stored.

Algorithm 3 demonstrates the definition of a *RegisterAction* named *reg_action*. The function simply adds 1 to the incoming value.

Algorithm 3: Definition of RegisterAction

---

1 RegisterAction<data_t, index_t, out_t>(target_array) reg_action = {
2   void apply(inout data_t value, out out_t return_value) {
3   *value* ← *value* + 1;
4   *return_value* ← *value*;
5   }
6 };
7 *rv* ← test_reg_action.execute(*index*);

6.2 Deploy Register Arrays and Relevant Actions

In the data plane of a PISA-based programmable switch, the processing of packets in the pipeline is divided into multiple stages. Dependent acts must be separated into stages. Packets can traverse the pipeline with global variables. To deploy the register arrays and relevant actions in the pipeline under the constraints of PISA, this paper takes into account VotePipe's workflow and the relationship between the three register arrays $<key, age>$, v_y and v_n.

As shown in Fig. 6, the three register arrays and their associated actions are placed in three stages i, $i + 1$ and $i + 2$ ($1 \leq i \leq n - 2$, n being the maximum number of stages). Assuming a packet enters the processing logic of VotePipe at stage i. An action is chosen in stage i based on the logic from line 1 to line 15 in Algorithm 1 to access or update the register array for $<key, age>$ values. If the packet is not discarded, it proceeds to stage $i + 1$ along with global variables *matched* and *age*. In stage $i + 1$, VotePipe performs the algorithm logic from line 16 to line 28 to update the register array for v_y based on the values of *matched* and *age*. Afterward, the packet and the variable v_y enter stage $i + 2$ together and execute the algorithm logic from line 29 to line 35 in stage $i + 2$. We present a self-defined header named *vote_pipe_h*, which consists of three members: *index*, *cKey*, and *cYes*. When a recirculation operation is triggered in stage $i + 2$, the packet is cloned, and some useful information will be stored in the header field for loopback and updating.

7 Evaluations

7.1 Experimental Setup

In this paper, we use a testing platform including a server, a traffic generator, and a programmable switch. The topology is depicted in Fig. 7. The devices are connected by two full-duplex links with a maximum rate of 40Gbps. The server is powered by an Intel Xeon E5-2699V4 @2.20GHz CPU based on the X86_64 architecture, featuring 22 cores and 44 threads, and running the Ubuntu 18.04 LTS operating system. The traffic generator Ixia can customize the patterns of traffic to be generated, such as protocol, packet size, and transmission mode, and generate traffic flows at ultra-high speeds (up to 400Gbps per port) to perform testing tasks. The programmable switch used in this study is the EmbedWay BaiSwitch8120, which is a hardware programmable switch based on Intel Tofino ASIC. The switch's data plane is based on PISA, with 32 ports capable of a maximum speed of 100Gbps and a maximum throughput of 3.2Tbps. It has two programmable pipelines, each with 12 stages, and each stage is allocated 1.5 MB of SRAM resources for shared use by all functions. The control plane is a x86 desktop running the Ubuntu16.04 LTS operating system, with an Intel Atom C3538 @2.20GHz CPU with 4 cores. The data plane and control plane of the switch are interconnected by a PCIe3.0 bus.

In the experiments, we use five real-world network traffic datasets from the WIDE [30] project, which captures the network traffic from a backbone network connecting Japan and the United States across various time periods. The selected sampling point collects 900 s of traffic data from a WIDE link with a 1Gbps rate to an upstream

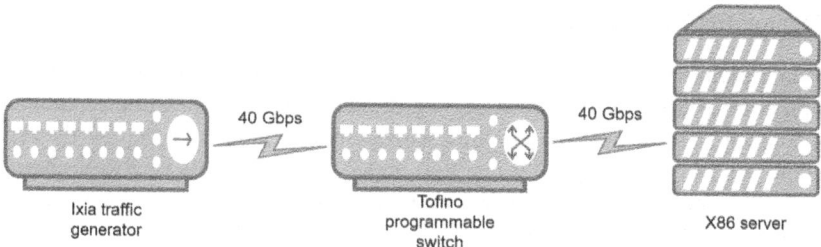

Fig. 7. Testing Platform

ISP. The collection period is from 14:00:00 to 14:15:00 every day. In this paper, five data sets (May 1, May 2, May 5, May 14, and May 21) are selected at random from the traffic data of May 2022, and 100,000 consecutive data packets are randomly chosen from each of the five data sets for analysis.

7.2 Baseline Algorithms and Metrics

Although there are many HHD algorithms implemented on P4 software programmable switches, only a few can be deployed on P4 hardware programmable switches. This study argues that algorithms that can only be implemented on software switches are of little practical use because of the poor performance of software switches. Given this, this study contrasts three hardware programmable switch algorithms: the traditional Count-min Sketch and the most recent PRECISION and dSketch. To effectively utilize the SRAM resources, it is best to set the size of the register array to a power of 2. Thus, for this experiment, k is set to either 128 or 512, indicating the detection of 128 or 512 large flows. The CRC_16 algorithm is used as the hash function. It can generate multiple hash functions by setting different seeds. Different algorithms' storage units require distinct register units to be implemented, hence the following configuration is adopted:

- **Count-min Sketch:** 3 columns, each with 128 or 512 counters. Each counter has a width of 32 bits.
- **PRECISION:** 3 stages, each with 128 or 512 entries. The flow identifier *key* and counter *val* have a width of 32 bits.
- **dSketch:** 3 columns, each with 128 or 512 counters and corresponding timestamps. Both the counter and the timestamp have a width of 32 bits.
- **VotePipe:** 1 table with 128 or 512 entries, where each entry comprises 4 fields, each with a width of 32 bits.

The experiments focus on the following performance metrics:

- **Accuracy rate**: the proportion of large flows detected by the algorithm among all flows.
- **Recall rate**: the proportion of large flows detected by the algorithm among all actual large flows.

- **False positive rate**: the proportion of non-large flows incorrectly detected as large flows by the algorithm among all non- large flows.
- **False negative rate**: the proportion of actual large flows not detected by the algorithm among all actual large flows.
- **Throughput**: comparison of the throughput of different algorithms under a 40 Gbps traffic rate.
- **Overhead**: comparison of the memory usage and loopback packet ratio of different algorithms.

7.3 Experimental Results

Accuracy. Accuracy represents the ability of an algorithm to distinguish between large flows and non-large flows. In the accuracy test, the large flows reported by each

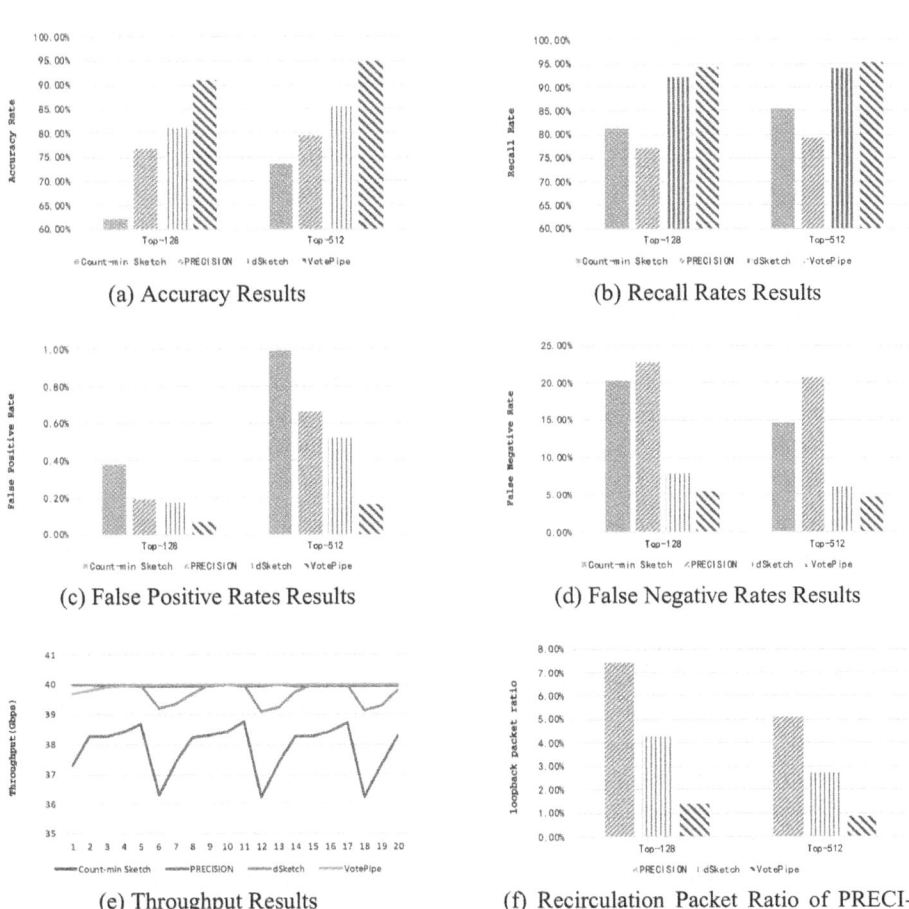

(a) Accuracy Results

(b) Recall Rates Results

(c) False Positive Rates Results

(d) False Negative Rates Results

(e) Throughput Results

(f) Recirculation Packet Ratio of PRECISION, dSketch and VotePipe

Fig. 8. Experimental Results

algorithm are compared with the largest 128 or 512 flows in the dataset, and the ratio of the number of detected large flows to the total number of reported flows is calculated. The results of the 5 datasets are averaged to obtain the final result, shown in Fig. 8a. The figure indicates that a larger value of k (512) generally results in higher accuracy, as it reduces hash collisions, estimating the count value more accurately. The Count-min Sketch's accuracy is low since each non-large flow packet increases the corresponding counter value, resulting in overestimation for each flow. The accuracy of PRECISION is influenced by two factors: the period reset mechanism and the minimum-value replacement strategy, which may result in the replacement of potentially large flows. With the addition of the time decay mechanism, dSketch achieved 4.39% (top-128) and 6.10% (top-512) greater accuracy than PRECISION. VotePipe, benefiting from both the time decay mechanism and the voting mechanism, addresses the problem of overestimation in Count-min Sketch. It also increases the probability that potential large flows will stay in the table, contributing to a significant improvement in accuracy. For top-128 flows and top-512 flows, the accuracy of VotePipe reached 91.09% and 94.83%, respectively, with a performance improvement ranging from 5.63% to 21.17% compared to the other three algorithms.

Recall Rates. Recall is an indicator of an algorithm's ability to accurately detect large flows. In the recall test, the large flows reported by each algorithm are compared to the largest 128 or 512 flows in the dataset, and the ratio of the number of detected large flows to the largest 128 or 512 flows is calculated. The average results of the 5 datasets are presented in Fig. 8b. As shown in the figure, sketch-based algorithms (Count-min Sketch and dSketch) exhibit higher recall rates, because sketch generally results in an overestimation of flow count values, making it easier to meet the conditions of large flows. PRECISION reports fewer large flows due to probabilistic recirculation and period reset mechanisms. dSketch, which utilizes a count value decay mechanism, reports fewer large flows than Count-min Sketch and has a higher recall rate. VotePipe, benefiting from the dynamic flow decay mechanism and the voting mechanism, achieves the highest recall rate of 94.40% and 95.27%.

False Positive Rates. The false positive rate represents the ratio of non-large flows identified as large flows in the detection results of an algorithm. It indicates the likelihood of non-large flows being incorrectly recognized as large flows. The false positive rates of each algorithm are shown in Fig. 8c. It is evident from the figure that all algorithms have very low false positive rates, with the highest being only 1%. All algorithms are more likely to report non-large flows as large flows as the number of target heavy hitters to be detected rises (from 128 to 512), although the rise in the false positive rate is still small. Theoretically, VotePipe is immune to reporting non-large flows as large flows because report operation is only triggered when positive votes reach the threshold T. However, the initial positive vote count of a replaced flow is set to half of the original negative vote count when the update operation is performed, which may lead to an overestimation of the count values of non-large flows and result in false reporting. Nonetheless, compared to VotePipe's high accuracy and recall rates, the false positive rate caused by the positive vote's initialization during flow updates is still acceptable. Furthermore, VotePipe still exhibits lower false positive rates compared to other algorithms.

False Negative. The false negative rate represents the ratio of large flows not detected by the algorithm to the total number of large flows, indicating the likelihood of the algorithm missing large flows. Figure 8d displays the false negative rates for several algorithms. The Count-min Sketch and PRECISION algorithms, which employ periodic reset, have greater false negative rates. This is due to the fact that periodic reset causes large flows that straddle reset time points to have their count values understated. On the other hand, dSketch with a counting decay mechanism over time and VotePipe with a counting decay mechanism over "flow age" are less likely to overlook significant flows that span periods, leading to reduced false negative rates.

Throughput. The throughput performance of the four algorithms is evaluated using the traffic generator Ixia at a rate of 40Gbps. Although Ixia is capable of generating traffic at a very high rate, it can only set relatively coarse-grained traffic patterns and lacks the multidimensional information of PCAP files from real datasets. Therefore, the following settings are applied for throughput testing with Ixia: all packets are set to be IPv4 protocol packets with a size of 64B. The destination IP address is used as the flow identifier, and 50,000 known large flows and 950,000 random non-large flows are set. The sending mode is continuous. Each sending cycle consists of 1,000 packets, including 700 large flow packets and 300 non-large flow packets. The sending rate is set to 40Gbps. As for the algorithms, the number of hash table entries is set to 65,536, which can cover the *top-65,536* flows among 1 million flows. The large flow threshold T is set to 300. The reset and decay periods are set to 6 s, which is suitable for Count-min Sketch, PRECISION, and dSketch. For VotePipe, age_t is set to 200. The algorithms are deployed on the switch, with the exit port still set as Ixia, which allows for real-time observation of throughput on Ixia. The throughput of each algorithm is recorded every second for 20 s. The results are presented in Fig. 8e.

The figure indicates that all algorithms can maintain good throughput in a fixed traffic pattern. Count-min Sketch can maintain full throughput throughout the process because it does not require loopback packets, but it has poor precision and cannot store flow identifiers. The throughput of PRECISION fluctuates with the reset period. This is because the probability of loopback packets decreases as the minimum count value in the table increases. When a cycle ends, the count values are reset to 0, and the process is repeated. dSketch is based on count decay over time, and its throughput only drops slightly at the beginning of each decay period. VotePipe experiences some packet replacement at the beginning, but after the count values stabilize, VotePipe can maintain maximum throughput.

Overhead. To be applicable in actual networks with massive traffic, this paper computes the SRAM resources and the number of data plane stages utilized by each algorithm when identifying 65,536 heavy hitters, as presented in Table 1. The hardware resources in the pipeline are shared by the basic functions of the data plane and other customized algorithms running. As shown in the chart, although the Count-min Sketch uses very few SRAM resources, it requires six pipeline stages to support a scale of $d = 3$. The number of stages needed increases rapidly as d increases due to the large number of comparisons required to determine the minimum value. As these comparisons are dependent, each will take up a stage. As a result, implementing more algorithms becomes nearly difficult. PRECISION requires more SRAM resources than Count-min

Table 1. Utilized SRAM resources and stages of the four algorithms ($k = 65,536$, other configurations see Sect. 7.2).

Algorithm	Utilized SRAM (MB)	Percentage of Utilized SRAM	Utilized Stages
Count-min Sketch	0.75	5.0%	6
PRECISION	1.5	10.0%	11
dSketch	1.5	10.0%	6
VotePipe	2	13.3%	4

Sketch as it needs to store flow identifiers. Even though PRECISION can apply the idea of stage folding, which involves placing the *val* field of the first sub-table and the *key* field of the second sub-table in the same stage, to decrease the number of occupied stages, its probability calculation operation almost fills up all the pipeline stages. dSketch employs a sketch structure, thereby occupying the same number of stages as Count-min Sketch. Also, dSketch necessitates storing timestamp information, which demands more SRAM resources. Among the four algorithms, VotePipe has the most recorded fields and hence consumes the most SRAM resources in proportion. Nevertheless, it only accounts for 13.3% of the total SRAM resources in the pipeline. Furthermore, VotePipe only uses four stages, which is the least among the four algorithms, and thus has the smallest impact on the pipeline processing speed.

Next, we evaluate the overhead of the packet recirculation mechanism in dSketch, PRECISION, and VotePipe. In this study, a separate register with a width of 32 bits and an initial value of 0 is set at the entrance of the pipeline for each algorithm. Whenever a looped packet arrives, the value of the register is incremented by 1. We replayed five WIDE datasets and calculated the ratio of looped packets for each algorithm, resulting in the average packet looping rate for each algorithm as shown in Fig. 8f. As can be seen, PRECISION and dSketch have a relatively high number of looped packets. PRECISION uses the inverse of the minimum value among d table entries as the recirculation probability, hence when network traffic is heavy, it is likely to experience lots of looped packets due to a large number of hash collisions. dSketch requires a loop if the time difference in any of the d stages exceeds the preset threshold, also resulting in many loop packets. VotePipe filters most non-large flows with the negative vote v_n, and loops are triggered only when the replacement condition is met. Consequently, in comparison to dSketch and PRECISION, VotePipe shows a great reduction in the number of loop packets, up to 74.7%.

In summary, the results show that VotePipe successfully meets the four design goals.

8 Conclusion

To tackle the challenge of deploying heavy hitter detection algorithms in the data plane of programmable switches, this paper proposes VotePipe, which utilizes a "flow age" based flow count decay mechanism and a voting-based flow filtering and updating mechanism. We evaluate the performance of VotePipe and other algorithms deployable on hardware programmable switches in terms of accuracy, throughput, memory overheads, and loopback packet ratio. The results show that VotePipe achieves higher accuracy than other algorithms while maintaining high throughput under acceptable overhead.

Acknowledgment. The work was supported by the National Natural Science Foundation of China under Grant No. 61972101.

References

1. Explore the power of intel® intelligent fabric processors. https://www.intel.com/content/www/us/en/products/network-io/programmable-ethernet-switch.html
2. Intel® tofino™ series programmable ethernet switch ASIC. https://www.intel.com/content/www/us/en/products/network-io/programmable-ethernet-switch/tofino-series.html
3. Cisco IOS netflow (2017). https://www.cisco.com/c/en/us/products/ios-nx-os-software/ios-netflow/index.html
4. Ben-Basat, R., Chen, X., Einziger, G., Rottenstreich, O.: Efficient measurement on programmable switches using probabilistic recirculation. In: 2018 IEEE 26th International Conference on Network Protocols (ICNP), pp. 313–323. IEEE (2018)
5. Ben-Basat, R., Einziger, G., Friedman, R., Kassner, Y.: Heavy hitters in streams and sliding windows. In: IEEE INFOCOM 2016 - The 35th Annual IEEE International Conference on Computer Communications, pp. 1–9 (2016). https://doi.org/10.1109/INFOCOM.2016.7524364
6. Benson, T., Akella, A., Maltz, D.A.: Network traffic characteristics of data centers in the wild. In: Proceedings of the 10th ACM SIGCOMM Conference on Internet Measurement, pp. 267–280 (2010)
7. Benson, T., Anand, A., Akella, A., Zhang, M.: Microte: fine grained traffic engineering for data centers. In: Proceedings of the Seventh Conference on Emerging Networking Experiments and Technologies, pp. 1–12 (2011)
8. Bosshart, P., et al.: P4: programming protocol-independent packet processors. ACM SIGCOMM Comput. Commun. Rev. **44**(3), 87–95 (2014)
9. Chiesa, M., Rétvári, G., Schapira, M.: Lying your way to better traffic engineering. In: Proceedings of the 12th International on Conference on emerging Networking Experiments and Technologies, pp. 391–398 (2016)
10. Chowdhury, S.R., Bari, M.F., Ahmed, R., Boutaba, R.: Payless: a low cost network monitoring framework for software defined networks. In: 2014 IEEE Network Operations and Management Symposium (NOMS), pp. 1–9 (2014). https://doi.org/10.1109/NOMS.2014.6838227
11. Cormode, G., Muthukrishnan, S.: An improved data stream summary: the count-min sketch and its applications. J. Algorithms **55**(1), 58–75 (2005)
12. González, L.A.Q., Castanheira, L., Marques, J.A., Schaeffer-Filho, A., Gaspary, L.P.: Bungee: an adaptive pushback mechanism for DDoS detection and mitigation in P4 data planes. In: 2021 IFIP/IEEE International Symposium on Integrated Network Management (IM), pp. 393–401 (2021)
13. Katta, N., et al.: Clove: congestion-aware load balancing at the virtual edge. In: Proceedings of the 13th International Conference on emerging Networking Experiments and Technologies, pp. 323–335 (2017)
14. Khooi, X.Z., Csikor, L., Li, J., Kang, M.S., Divakaran, D.M.: Revisiting heavy-hitter detection on commodity programmable switches. In: 2021 IEEE 7th International Conference on Network Softwarization (NetSoft), pp. 79–87 (2021). https://doi.org/10.1109/NetSoft51509.2021.9492531

15. Khooi, X.Z., Csikor, L., Li, J., Kang, M.S., Divakaran, D.M.: Revisiting heavy-hitter detection on commodity programmable switches. In: 2021 IEEE 7th International Conference on Network Softwarization (NetSoft). pp. 79–87 (2021). https://doi.org/10.1109/NetSoft51509.2021.9492531
16. Lakhina, A., Crovella, M., Diot, C.: Characterization of network-wide anomalies in traffic flows. In: Proceedings of the 4th ACM SIGCOMM Conference on Internet Measurement, pp. 201–206 (2004)
17. Lei, Y., Yu, L., Liu, V., Xu, M.: Printqueue: performance diagnosis via queue measurement in the data plane. In: Proceedings of the ACM SIGCOMM 2022 Conference, pp. 516–529 (2022)
18. Li, Y., Miao, R., Kim, C., Yu, M.: {FlowRadar}: a better {NetFlow} for data centers. In: 13th USENIX Symposium on Networked Systems Design and Implementation (NSDI 2016), pp. 311–324 (2016)
19. Liu, Z., Manousis, A., Vorsanger, G., Sekar, V., Braverman, V.: One sketch to rule them all: rethinking network flow monitoring with univmon. In: Proceedings of the 2016 ACM SIGCOMM Conference, pp. 101–114 (2016)
20. Metwally, A., Agrawal, D., El Abbadi, A.: Efficient computation of frequent and top-k elements in data streams. In: Eiter, T., Libkin, L. (eds.) ICDT 2005. LNCS, vol. 3363, pp. 398–412. Springer, Heidelberg (2004). https://doi.org/10.1007/978-3-540-30570-5_27
21. Miao, R., Zeng, H., Kim, C., Lee, J., Yu, M.: Silkroad: making stateful layer-4 load balancing fast and cheap using switching ASICs. In: Proceedings of the Conference of the ACM Special Interest Group on Data Communication, pp. 15–28 (2017)
22. Moshref, M., Yu, M., Govindan, R., Vahdat, A.: Dream: dynamic resource allocation for software-defined measurement. In: Proceedings of the 2014 ACM Conference on SIGCOMM, pp. 419–430 (2014)
23. P4: P4 open source programming language. https://p4.org/
24. P4: P4 16 language specification (2022). https://p4.org/p4-spec/docs/P4-16-v-1.2.3.html
25. Qureshi, M.A., et al.: PLB: congestion signals are simple and effective for network load balancing. In: Proceedings of the ACM SIGCOMM 2022 Conference, pp. 207–218 (2022)
26. Ramachandran, A., Seetharaman, S., Feamster, N., Vazirani, V.: Fast monitoring of traffic subpopulations. In: Proceedings of the 8th ACM SIGCOMM Conference on Internet Measurement, IMC 2008, pp. 257-270. Association for Computing Machinery, New York (2008). https://doi.org/10.1145/1452520.1452551
27. sFlow: Sampling rates. https://blog.sflow.com/2009/06/sampling-rates.html
28. Sivaraman, V., Narayana, S., Rottenstreich, O., Muthukrishnan, S., Rexford, J.: Heavy-hitter detection entirely in the data plane. In: Proceedings of the Symposium on SDN Research, pp. 164–176 (2017)
29. Turkovic, B., Oostenbrink, J., Kuipers, F., Keslassy, I., Orda, A.: Sequential zeroing: online heavy-hitter detection on programmable hardware. In: 2020 IFIP Networking Conference (Networking), pp. 422–430. IEEE (2020)
30. WIDE: Wide. https://www.wide.ad.jp/
31. Yang, T., et al.: Elastic sketch: adaptive and fast network-wide measurements. In: Proceedings of the 2018 Conference of the ACM Special Interest Group on Data Communication, pp. 561–575 (2018)
32. Yang, T., et al.: Heavykeeper: an accurate algorithm for finding top-k elephant flows. IEEE/ACM Trans. Networking **27**(5), 1845–1858 (2019). https://doi.org/10.1109/TNET.2019.2933868
33. Zhou, Y., et al.: Cold filter: a meta-framework for faster and more accurate stream processing. In: Proceedings of the 2018 International Conference on Management of Data, SIGMOD 2018, pp. 741–756. Association for Computing Machinery, New York (2018). https://doi.org/10.1145/3183713.3183726

SDN Based Network Path Planning Optimization for Printing Cloud Service

Jiajun Peng[1], Qian He[1(✉)], Qi Pan[1], and Yanbo Liu[2]

[1] Guangxi Key Laboratory of Cryptograph and Information Security, Guilin University of Electronic Technology, Guilin 541004, China
heqian@guet.edu.cn
[2] Zhuhai Xinye Electronic Technology Co., Ltd., Zhuhai 519075, China
zxywliu@xprinter.net

Abstract. With the rapid development of information technology and the widespread application of cloud computing, cloud printing services have become an efficient and convenient printing solution, widely used in modern offices and corporate environments. However, as the scale of cloud printing services expands and users' expectations for service quality increase, the system faces several challenges. Among them, monitoring the status of resources such as computing, storage, and application services, as well as reliable data transmission path planning, have become critical issues. In response to these challenges, this paper aims to provide intelligent printing terminal routing strategies for Zhuhai Xinye Electronic Technology Company, ensuring comprehensive reliability for different levels of business transmission. This study proposes a QoS-aware path planning mechanism based on Software-Defined Networking (SDN) to address the prioritization and packet loss issues during the massive real-time data transmission in cloud-edge collaboration. The mechanism allows flexible adjustment of network resource demands and priorities for different levels of business through SDN. Additionally, it incorporates blockchain technology to store path planning and alert issuance flow table information, ensuring the traceability of business path and alert information.

Keywords: Cloud-edge collaboration · SDN · Blockchain · Network resource optimization

1 Introduction

With the rapid development and widespread adoption of information technology and cloud computing, cloud services have become essential infrastructure in modern society. Among various cloud applications [1], cloud printing services stand out as efficient and convenient printing solutions, widely used in office environments, schools, and businesses. By submitting print tasks to the cloud, users can access printing services from any device, enabling cross-platform and cross-regional printing capabilities, significantly enhancing office efficiency and convenience.

However, as the scale of cloud printing services continues to expand and user demands for service quality increase, the system faces a series of challenges. Proper allocation of network resources and effective path planning become crucial issues for cloud printing services. In the cloud computing environment, different types of businesses have varied requirements for network transmission, and traditional network path planning algorithms often fail to meet the diverse business needs. Particularly with the extensive use of new network technologies like Software-Defined Networking (SDN), traditional network vertical integration is disrupted, separating network control logic from underlying routers and switches [2], thus demanding more efficient and intelligent network management and resource allocation for the cloud printing system.

Furthermore, data transmission in cloud printing systems involves a significant amount of real-time data, and problems such as packet loss and network congestion can lead to a decline in user experience and service quality [3]. Traditional network path planning algorithms often struggle to adapt flexibly to changing business needs and network conditions. To ensure the cloud printing system provides services efficiently and reliably, there is a need to introduce new path planning mechanisms under SDN networks to meet the transmission demands of different business types and improve network resource utilization and service quality.

In recent years, scholars both domestically and internationally have conducted extensive research to guarantee reliable transmission of different-level businesses in SDN networks and enhance network service quality. Foreign researchers, such as Amir and colleagues, have proposed a series of solutions for rational network resource allocation and path planning to improve network service quality. For example, they designed the SDN-MPLS algorithm, optimizing network services by balancing network load balancing, routing length, energy efficiency, and complexity between mobile networks.

Researchers have also focused on improving network transmission reliability and resource utilization. One study [5] proposed a potential traffic bottleneck cutting and identification scheme, significantly improving network service quality through random edge path selection, shortest path edge selection, and edge capacity selection methods. Additionally, another study introduced [6] the QCORS SDN routing algorithm, which uses flexible SDN technology to meet the diverse communication requirements in smart grids. QCORS predicts future link congestion states and constructs a virtual non-congested topology to route data packets through lower load links, leading to reduced average delay and packet loss compared to other methods.

Moreover, an efficient policy-based hybrid SDN network routing method was proposed [7], utilizing idle IP addresses to save flow table space and providing traffic engineering support. The performance of this method was found to be superior to other controllers. Regarding SDN-based mobile network traffic engineering, a new optimization routing algorithm called PSLC was presented [4]. Its objectives are to improve network resource utilization, load balancing, and reduce routing costs and complexity.

Within China, a novel data center network traffic scheduling mechanism combining SDN and swarm intelligence algorithms was proposed [8]. Through simulation verification on the Mininet platform, the method significantly improved the average bandwidth of subcarrier pairs compared to ECMP and ACO-SDN algorithms, while reducing the maximum link utilization.

Additionally, a SDN-based multi-tenant network resource allocation and management platform was designed [9]. This platform provides different view interfaces for different tenant networks, realizing resource allocation and traffic isolation. The method also utilizes idle path forwarding in links to achieve energy-saving objectives. Xu Bo and colleagues [9] proposed a routing optimization method based on deep deterministic policy gradients, combining reinforcement learning and SDN (DDPG) models in the programmable data plane. This method aims to address routing decision biases caused by uneven traffic distribution and inaccurate network state measurements in data center networks.

However, despite the existing research providing some guidance for SDN network path planning and resource allocation, the uniqueness of cloud printing services and the increasing user demands call for more intelligent and flexible solutions. To address this, this paper proposes a cloud-edge collaborative cloud printing system architecture, aiming to fully leverage the advantages of cloud computing and edge computing, optimize network resource allocation and path planning, and enhance the performance and reliability of cloud printing services.

Within this architecture, this study introduces a QoS-aware path planning assurance mechanism based on SDN. This mechanism dynamically selects the optimal transmission path based on the requirements of different-level businesses and packet loss issues, thereby improving the reliability and efficiency of data transmission. Additionally, to ensure the security and integrity of path planning and alert information, this paper integrates blockchain technology to achieve trusted storage of path and alert information, ensuring the security of the data.

The main contributions of this paper are as follows:

Proposing a cloud-edge collaborative cloud printing system architecture that combines cloud computing and edge computing to optimize print task processing and transmission, enhancing system performance and reliability.

(1) Introducing a QoS-aware path planning guarantee mechanism based on SDN, dynamically selecting the optimal transmission path according to business priorities and packet loss issues to enhance data transmission reliability and efficiency.
(2) Utilizing blockchain technology to ensure the trustworthy storage of path information and alert information, guaranteeing data security and integrity.

The rest of this paper is organized as follows. Section 3 presents the SDN based Network Path Planning Optimization, which includes four parts: the cloud-edge collaborative cloud printing system architecture, the design of smart contracts, the QoS-aware path planning algorithm, and the link state alerting mechanism. Section 4 describes the experimental evaluation of the proposed method. Finally, Sect. 5 provides the conclusion and outlook of this paper.

2 System Modeling

To enhance the cloud service quality of printing terminal devices, this study proposes a cloud-edge collaborative cloud printing system architecture, as depicted in Fig. 1. The architecture comprises two main layers: the Cloud Center Layer and the Edge Node Layer.

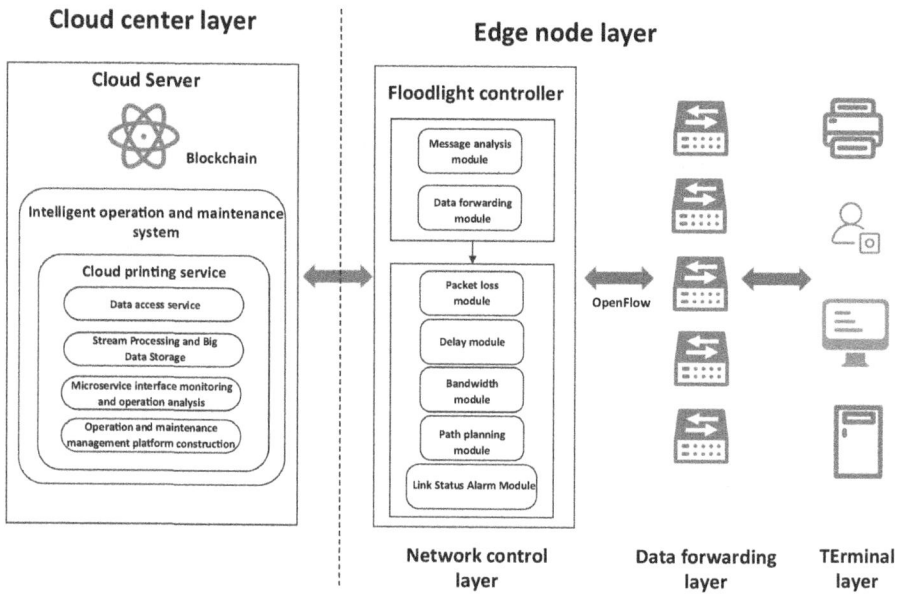

Fig. 1. Overall system architecture diagram

Cloud Center Layer: In this layer, we deploy blockchain services and an intelligent operations and maintenance (O&M) system to support the secure and efficient operation of the cloud printing system.

Blockchain Services: Blockchain is a distributed and tamper-resistant data storage and transmission technology commonly used to record transaction information and data history. In our cloud-edge collaborative cloud printing system architecture, blockchain services are employed to securely and reliably store path planning information and link state alerting information. Leveraging the characteristics of blockchain technology, we ensure the immutability of path planning and alerting information throughout the entire system, guarding against malicious tampering or forgery of information.

Intelligent O&M System: The intelligent O&M system plays a pivotal role in the cloud-edge collaborative cloud printing system. Its primary responsibility is to monitor and collect the status and operational data from various devices in the network, such as printers, PCs, and servers. As the central controller of SDN, the Floodlight controller at the edge can monitor and collect various information related to network devices, such as traffic, link states, and device connections. This data is then transmitted through the SDN network to the intelligent O&M system on the cloud server. The intelligent O&M system's functionality extends beyond data collection; it also analyzes and processes the collected data. For instance, by monitoring packet loss rates, bandwidth utilization, and latency of links, the intelligent O&M system can assess the current network status and performance, aiding the decision-making process of the path planning algorithm. Additionally, the intelligent O&M system conducts fault detection and prediction, promptly identifying abnormal situations in the network and taking corresponding measures to enhance network reliability and stability.

Edge Node Layer: The Edge Node Layer consists of three components: Network Control Layer, Data Forwarding Layer, and Terminal Layer. These layers collaborate to support the efficient operation of the cloud printing system.

Network Control Layer: At the core of this layer is the Floodlight controller, which achieves centralized management and control of the entire network, monitoring and adjusting network behavior. The Floodlight controller dynamically configures traffic, routing, and topology structures, enabling network flexibility and optimization. Through the OpenFlow protocol, it communicates with the OpenFlow switches in the Data Forwarding Layer to jointly implement network control.

Data Forwarding Layer: This layer includes P4 forwarding devices and OpenFlow switches, responsible for packet processing and forwarding tasks. During data forwarding, data packets receive corresponding forwarding decisions based on routing tables, flow tables, and other rules, and are sent to the destination address. This design ensures efficient data transmission and processing in the network, providing stable printing services to users.

Terminal Layer: As a critical component of user network access, the Terminal Layer comprises hosts (e.g., personal computers, laptops, servers), switches, and mobile devices. The Terminal Layer sends IP packets containing the required attribute set for access requests to P4 forwarding devices and conveys the relevant information through filling in the Options field of the data packet. This way, users can conveniently access cloud printing services, enabling efficient printing functionality.

3 Proposed Optimization Method(s)

3.1 Smart Contract

To ensure the traceability of path planning and link status alert information, this study introduces smart contracts to design the storage and retrieval of path and alert status information.

1. Path Planning Contract (PathRoute): A path planning information structure is designed, comprising the source and destination terminal MAC addresses (mac) and the allocated path information (path). The PathRoute contract includes two crucial methods: AddPathRoute() and QueryHostPathRoute().

The AddPathRoute() method is employed to store the path information planned by the path planning module for different types of businesses on the blockchain. This method accepts input parameters and securely stores the path planning information in the smart contract, ensuring reliable storage of business path planning results for subsequent queries and usage by other relevant modules or terminals. The functionality code for adding path planning information in the contract is depicted in Algorithm 1.

Algorithm 1 AddPathRoute()

Input: mac, PathInfo
Output: success

1. **If** len(args) != 2 **then** {   //Verify the correctness of the incoming parameters
2.   return shim.Error("input args is illegal")
3. **End**
4. pathinfotest := flowinfo{
5.   mac : args[0],
6.   path : args[1],
7. }
8. Flow,_ := json.Marshal(pathinfotest)
9. APIstub.PutState(pathinfotest.mac,path)   // Store path information in the blockchain
10. **return** shim.Success(nil)

The QueryHostPathRoute() method is utilized to retrieve the planned path for a specific business based on the MAC addresses of the source and destination hosts. This method accepts the MAC addresses of the source and destination hosts as inputs and retrieves the relevant path planning information from the smart contract. The returned result is a structured data (pathinfo) containing detailed information about the planned path for the specific business. Through this method, other modules within the system or end-users can efficiently access the planned path information for specific businesses, facilitating further analysis and decision-making.

2. Alert Flow Table Contract (MyFlow): A data structure for alert flow table information is designed, consisting of the DPID (sid) of the switch to which the alert flow table is issued and the information about the alert flow table (flow). The MyFlow contract also includes two critical methods: AddFlow() and queryFlow().

The AddFlow() method is employed to securely store the information of the alert-issued flow table on the blockchain. This method receives input parameters, including relevant information about the alert flow table, and permanently stores this information in the smart contract. By calling the AddFlow() method, the information of the alert flow table is persistently stored to ensure its immutability and trustworthiness, and it can be queried and utilized by other relevant modules or end-users. The functionality code for adding alert flow table information in the contract is depicted in Algorithm 2.

Algorithm 2 AddFlow()

Input:	sid, flow	
Output:	success	
1.	**If** len(args) != 2 **then** {	//Verify the correctness of the incoming parameters
2.	return shim.Error("input args is illegal")	
3.	**End**	
4.	flowinfotest := flowinfo{	
5.	sid : args[0],	
6.	flow : args[1],	
7.	}	
8.	Flow,_ := json.Marshal(flowinfotest)	
9.	APIstub.PutState(flowinfotest.sid,flow)	// Store the alarm flow table information in the blockchain
10.	**return** shim.Success(nil)	

The queryFlow() method serves the purpose of retrieving alert flow table information for a specific switch based on its DPID (Data Path ID). This method takes the DPID of the switch as input and searches for the relevant alert flow table information within the smart contract. The returned result provides detailed information about the alert flow table associated with that particular switch, enabling other modules within the system or end-users to conduct analysis and utilize the data effectively.

Through the use of smart contracts, this research ensures the immutability and trustworthiness of both path planning and link status alert information. The storage mechanism of smart contracts guarantees the permanent storage of path planning information, while the query contract offers an efficient means to access the planned paths for specific businesses. Such design empowers various internal modules of the system to accurately obtain path planning information, enabling further analysis and decision-making, thereby enhancing the system's reliability and security.

3.2 QoS-Aware Path Planning Algorithm

Upon secure terminal access to the SDN network, upper-layer applications will utilize the connected terminals for transmitting different-level services. To meet the inter-terminal transmission requirements for diverse-level services, the SDN network's link bandwidth, delay, and packet loss rate parameters are harnessed for path planning. Furthermore, to ensure traceability of the business path planning results, the algorithmically planned paths are stored in the blockchain. This achieves the objective of maintaining dependable service transmission while enabling traceable business path information.

This algorithm calculates parameters such as link packet loss rate, bandwidth, and delay, and based on these calculations, it performs path planning for different-level services. To distinguish between service levels, the Type of Service (ToS) field in the matching domain of the OpenFlow protocol is utilized.

The specific design is as follows: When a terminal issues a command containing the ToS field, the connected OpenvSwitch switch receives the request. As the switch's cache lacks the relevant processing logic, it encapsulates the request into a Packet_in message and forwards it to the Floodlight controller. The controller parses the ToS and related parameters, and based on the ToS value, it executes the Business Grade Assurance Path Planning Algorithm for different-level services.

The path planning algorithm follows these steps:

① After the Floodlight controller is started, it employs the link discovery module to retrieve all paths in the network and stores them in the PathListsum collection.

② The controller checks if the current ToS value is 0. If it is, indicating a standard business type, the algorithm invokes the shortest path algorithm to allocate a path for this type of service.

③ For non-zero ToS values, the controller collects the packet loss rate of each link in the network every 60 s and stores the results in the DPI D_Loss_Map collection. Each key-value pair consists of a NodePortTuple object as the key and a Double-type PacketLoss value (PathPacketLoss) representing the overall packet loss rate of a specific link.

④ Once the packet loss rates for all links are collected, the controller iterates through the DPID_Loss_Map collection and compares each link's packet loss rate with the system's preset packet loss threshold. Links with packet loss rates below the threshold are added to the PathList collection for subsequent service type selection. Links with packet loss rates above the threshold are not included in the PathList collection.

The pseudocode for steps 1 to 4 is presented in Algorithm 3 below.

Algorithm 3

Input:	Bandwidth$_{available}$, LinkLoss, ToS, Link$_{delay}$, Bandwidth$_{link}$, Linkct$_{delay}$
Output:	none

1. Floodlight ← OF switch Packet_In
2. ToS ← Floodlight parses the Packet_In message to get the ToS value
3. List<path> pathListsum □ The controller obtains all paths in the network
4. **if** ToS == 0 **Then**
5.   return Path;   //return shortest path
6. **End**
7. DPID_Loss_Map<NodePortTuple,Double> ← Call the packet loss rate measurement module to detect link packet loss
8. **if** (PathListsum.size()==1) **then**
9.   return PathListsum.get(0)   //return this unique path
10. **else if** PathListsum == Ø **then**
11.   return null;
12. **End if**
13. **End if**
14. **For** each i ∈ [0, PathListsum.size()] **do**
15.   **For** each NodePortTuple(np) □ switchPorts **do**
16.   Double ls = DPID_Loss_Map.get(np)
17.   Packetloss = ls + Packetloss   //Calculate the packet loss rate of each link
18.   **End for**
19.   **If** Packetloss < LinkLoss **Then**
20.   PathList.add(PathListsum.get(i))   //Add links that meet the packet loss default value to the PathList collection
21.   **End**
22. **End for**

⑤ The algorithm evaluates whether multiple paths exist in the PathLB collection. If there are multiple paths, it sorts them in descending order based on available bandwidth and selects the first path for allocation.

⑥ If there is only one path in the PathLB collection, the algorithm uses the K-shortest path algorithm to traverse the paths in the PathList collection. It then sorts these paths in ascending order based on the number of hops and stores them in the PathListHop collection. The algorithm then evaluates the current service's ToS value. For ToS = 8, the algorithm prioritizes paths with maximum available bandwidth. For ToS = 16, it prioritizes paths with the lowest delay. For ToS = 0, it selects the path with the fewest hops.

The pseudocode for steps 5 to 6 is presented in Algorithm 4 below.

Algorithm 4

Input:	Bandwidth$_{available}$, LinkLoss, ToS, Link$_{delay}$, Bandwidth$_{link}$, Linkct$_{delay}$
Output:	path

1. **If** ToS == 8 **then**   //Services that require high bandwidth
2.   **For** each i ∈ [0, pathList.size()]   **do**
3.   **if** (linkctdelay<=linksdelay &&Bandwidthavailable>=Bandwidthbus)   **Then**
4.   PathLB.add(PathList.get(i))   //Add paths that satisfy the conditions to PathLB
5.   **End**
6.   **End for**
7.   **If** PathLB.size > 0   **then**
8.   **For** each i ∈ [0, PathLB.size()]   **do**
9.   PathLB ← 降序排序
10.   Path ← PathLB.get(0)   //Get the path with the largest available bandwidth
11.   **End for**
12.   **Else if** PathLB.size <= 0   **then**
13.   **For** each i ∈[0,PathList.size()]   **do**
14.   PathList ← k shortest path algorithm, sorted by hop count from low to high
15.   PathListHop = PathList;   //Put the selected K shortest paths into
16.   **If** ToS == 8   **then**
17.   **For** each i ∈ [0, PathListHop.size()]   **do**
18.   Path ← Select the path allocation with the largest bandwidth
19.   **End for**
20.   **Else if**   ToS == 16   **then**
21.   **For**   each i ∈ [0, PathListHop.size()]   **do**
22.   Path ←Select the path assignment with the least delay
23.   **End for**
24.   **End if**
25.   **End for**
26.   **End if**
27. **End if**
28. **return** Path

⑦ For ToS = 16, the algorithm follows a similar workflow as for ToS = 8, with the additional criterion of selecting paths where the current path's bandwidth is less than or equal to the bandwidth threshold and the current service's link delay is less than or equal to the maximum delay threshold required by the service. Paths meeting these conditions are stored in the PathLD collection.

⑧ If multiple paths exist in the PathLD collection, the algorithm sorts them in ascending order based on delay and selects the first path for allocation. If there are no multiple paths, the algorithm reverts to step 6.

The pseudocode for steps 7 to 8 is presented in Algorithm 5 below.

Algorithm 5

Input: Bandwidth$_{available}$, LinkLoss, ToS, Link$_{delay}$, Bandwidth$_{link}$, Linkct$_{delay}$
Output: path

1. **if** ToS == 16 **then**
2.   **For** each i ∈ [0, PathList.size()] **do**
3.   **if** (linkctdelay<=linkdelay &&Bandwidthbus<=BandwidthLink)**Then**
4.   PathLD.add(PathList.get(i)) //Add paths that satisfy the conditions to PathLD
5.   **End**
6.   **End for**
7.   **If** PathLD.size > 0 **then**
8.   **For** each i ∈ [0, PathLD.size()] **do**
9.   PathLD ← sort ascending
10.   Path ← PathLD.get(0) //Get the path with the shortest delay
11.   **End for**
12.   **Else if** PathLD.size <= 0 **then**
13.   **For** each i ∈ [0, PathList.size()] **do**
14.   PathList ←k shortest path algorithm, sorted by hop count from low to high
15.   PathListHop = PathList; //Put the selected K shortest paths into
16.   **If** ToS == 8 **then**
17.   **For** each i ∈ [0,PathListHop.size()] **do**
18.   Path ← Select the path allocation with the largest bandwidth
19.   **End for**
20.   **Else if** ToS == 16 **then**
21.   **For** each i ∈ [0, PathListHop.size()] **do**
22.   Path ← Select the path assignment with the least delay
23.   **End for**
24.   **End if**
25.   **End if**
26.   **End if**
27. **End if**
28. **return** Path

⑨ After the algorithm completes, it executes the addPathRoute contract to store the pre-planned, business-level-based paths on the blockchain. This ensures the reliability and traceability of different-level service paths between endpoints.

3.3 Link State Alert Mechanism

The Business Grade Assurance Path Planning Algorithm is designed to ensure reliable transmission of different-level services between terminals. However, considering the variability of link states that may lead to service transmission unreliability, a Business Assurance Link State Alert Mechanism is devised to further ensure the dependable data transmission of services.

Firstly, the cost factors H, L, and B are established through the initialization of packet loss cost, delay cost, and bandwidth cost. Subsequently, the packet loss monitoring module monitors the current packet loss status of each link involved in service transmission.

If the current link's packet loss rate is greater than or equal to the packet loss threshold multiplied by H, a first-level alert is triggered. The controller determines the service type based on the link's ToS value and performs the following actions:

1. When ToS = 8 for the service, the controller first employs the delay measurement module to calculate the current link's delay status. If the link's delay exceeds or equals the maximum delay threshold required by the service multiplied by L, the available bandwidth of that link is further evaluated. If the service transmission's bandwidth surpasses the available bandwidth, a second-level alert is triggered. The controller issues an alert flow table to block the service transmission and records the alert flow table information in the blockchain through the alert information contract. Simultaneously, the controller re-executes the path planning algorithm to re-plan the transmission path for the current service. If both bandwidth and delay meet the service requirements, the current service transmission state remains unchanged.
2. When ToS = 16 for the service, the controller employs the link bandwidth measurement module to calculate the current link's bandwidth status. If the service transmission's bandwidth has reached the link's maximum bandwidth multiplied by B, the controller then utilizes the delay measurement module to calculate the transmission delay of that link. If the link's delay exceeds the required delay threshold for the service, a second-level alert is triggered. The controller takes the same actions as described in the previous scenario to handle the alert situation.

Through the design of the Business Assurance Link State Alert Mechanism, real-time link status monitoring is achieved, and alerts are determined based on preset thresholds and cost factors. The controller performs corresponding actions according to the alert triggers, including blocking service transmission, storing alert flow table information, and re-planning transmission paths. This ensures the reliable transmission of business data and leverages blockchain technology to achieve traceability and tamper resistance of alert information, enhancing the overall system's reliability and security. The mechanism integrates the Business Grade Assurance Path Planning Algorithm and Link State Alert Mechanism to provide comprehensive assurance for the dependable transmission of different-level services between terminals.

4 Experiment

In order to validate the functionality and superiority of the QoS-aware path planning algorithm, we designed a network topology using the Mininet simulation platform, as depicted in Fig. 2. This network comprises a Floodlight controller, 7 OpenvSwitch switches (S1–S7), and 4 hosts (Host 1–Host 4). By comparing the QoS-aware path planning algorithm with the equivalent multi-path ECMP routing algorithm and the shortest path algorithm in terms of packet loss rate, we can verify the superior performance of this algorithm.

In the same experimental environment, we conducted the following operations: employing the iperf tool, we designated Host 1 as the UDP server and Host 2, Host 3, and Host 4 as clients. Different Type of Service (ToS) commands were input into the terminals of Host 2, Host 3, and Host 4, and data transmission was initiated at 20-s

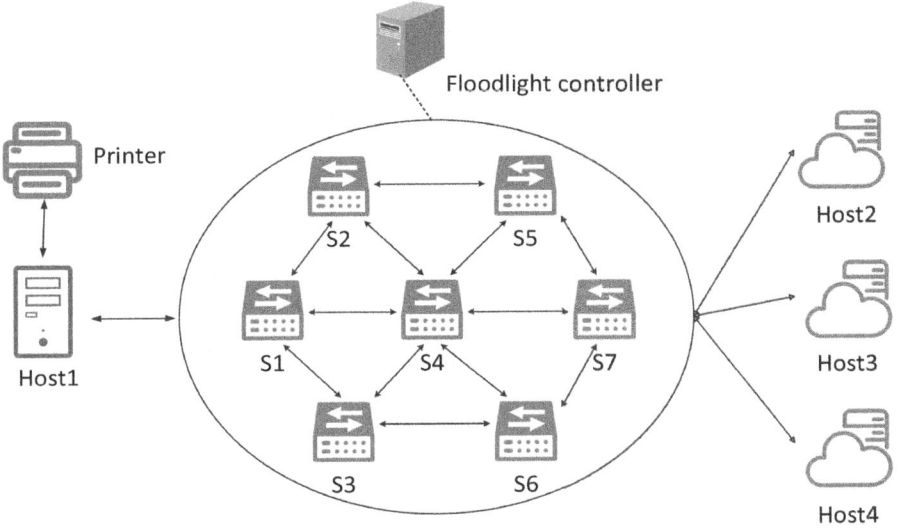

Fig. 2. Simulation experiment network topology

intervals. The initial link bandwidth was set to 5 Mbps, and the maximum bandwidth threshold for all links was set to 10 Mbps. The initial link delay was set to 45 ms, while the highest delay threshold required by the services was set to 40 ms. To assess the algorithm's performance, we randomly set the packet loss rate between S1–S4, S1–S2, and S6–S7 to 5%. Additionally, the packet loss rate threshold for the services was set to 5%.

We conducted 20 data transmissions for each type of service in the three algorithms, recording the packet loss rate for each algorithm under different types of services. Our analysis yielded the following results: under ToS = 0, ToS = 8, and ToS = 16, the average packet loss rate for our proposed approach was 7.22%, 1.74%, and 1.57% respectively. The equivalent multi-path ECMP algorithm had average packet loss rates of 17.79%, 20.55%, and 17.18%, while the shortest path algorithm exhibited average packet loss rates of 37.29%, 43.05%, and 41.96%. A comparative chart illustrating the packet loss rates for the three algorithms under different service types is shown in Fig. 3 below.

Through the above chart, we can intuitively compare the packet loss rate performance of the QoS-aware path planning algorithm, equivalent multi-path ECMP routing algorithm, and shortest path algorithm under different types of services. The experimental results demonstrate that our designed QoS-aware path planning algorithm outperforms the multi-path ECMP and shortest path algorithms in terms of packet loss rate. This provides robust technical support for optimizing and intelligently managing the cloud-edge collaborative cloud printing system architecture.

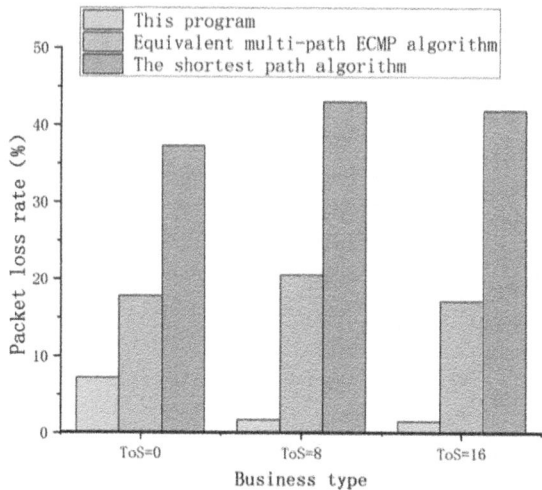

Fig. 3. Comparison chart of packet loss rates for three algorithms

5 Conclusion

This study delves into the key technologies and algorithms of the cloud-edge collaborative cloud printing system architecture and experimentally verifies the superiority of the proposed QoS-aware path planning assurance mechanism in terms of packet loss rate. The integration of cloud computing and edge computing through the cloud-edge collaborative architecture provides a more efficient and reliable solution for cloud printing services. Within this architecture, we employ an SDN-based QoS-aware path planning algorithm, combined with blockchain services for trustworthy storage of path and alert information, enhancing the system's reliability and security.

The results of this research demonstrate that the QoS-aware path planning algorithm exhibits low packet loss rates across various types of services. The algorithm flexibly selects suitable paths for data transmission based on business priorities and packet loss issues, effectively reducing packet loss and improving network service quality. Simultaneously, the application of blockchain technology enhances the credibility and tamper-proof nature of data transmission, ensuring the secure storage of path and alert information.

However, this research also faces certain limitations. In practical applications, the QoS-aware path planning algorithm still requires further optimization and improvement to meet complex network scenarios. Additionally, the efficiency of blockchain technology in storage and transmission needs further enhancement to cater to the demands of large-scale networks.

Future research directions include exploring the dynamic adaptability of path planning algorithms to cope with dynamic changes in network topology. Moreover, the incorporation of emerging technologies such as artificial intelligence and machine learning can further optimize path planning and resource allocation strategies.

Acknowledgment. This work is partly supported by the Natural Science Foundation of China (62162018), The Natural Science Foundation of Guangxi (No. 2019GXNSFGA245004) Funding.

References

1. Gupta, N., Sohal, A.: Cloud computing: evolution, research issues, and challenges. In: Emerging Computing Paradigms: Principles, Advances and Applications, pp. 1–17 (2022)
2. Kreutz, D., Ramos, F.M.V., Verissimo, P.E., et al.: Software-defined networking: a comprehensive survey. Proc. IEEE **103**(1), 14–76 (2014)
3. Li, W., Qi, H., Xu, R., et al.: Research progress and trend of data center network traffic scheduling. Chin. J. Comput. **43**(4) (2020)
4. Alidadi, A., Arab, S., Askari, T.: A novel optimized routing algorithm for QoS traffic engineering in SDN-based mobile networks. ICT Express **8**(1), 130–134 (2022)
5. Alzaben, N., Engels, D.W.: End-to-end routing algorithm based on max-flow min-cut in SDN controllers. In: 2022 24th International Conference on Advanced Communication Technology (ICACT), pp. 1–6. IEEE, Piscataway (2022)
6. Su, Y., Huan, C., et al.: A QoS-guaranteed and congestion-controlled SDN routing strategy for smart grid. Appl. Sci. **12**(15), 7629 (2022)
7. Paliwal, M., Nagwanshi, K.K.: Effective flow table space management using policy-based routing approach in hybrid SDN network. IEEE Access **10**, 59806–59820 (2022)
8. Dai, R., Li, H., Fu, X.: Data center traffic scheduling mechanism based on differential evolution and ant colony algorithm. Comput. Appl. **42**(12), 3863 (2022)
9. Pang, S., Zeng, D., Chen, X.: Research on SDN-based data center network traffic management and optimization. In: 2022 IEEE 2nd International Conference on Power, Electronics and Computer Applications (ICPECA), pp. 600–604. IEEE, Piscataway (2022)
10. Xu, B., Zhou, J., Wu, J., et al.: DDPG-based routing optimization method under programmable data plane. J. Comput. Eng. Appl. **58**(3) (2022)

A Survey on Security Issues of SDN Controllers

Rui Wang[1], Youhuizi Li[1], Meiting Xue[2](✉), Baokang Zhao[3](✉), Yuyu Yin[1], and Yangyang Li[1]

[1] College of Computer Science and Technology, Hangzhou Dianzi University, Hangzhou 310000, China
[2] School of Cyberspace, Hangzhou Dianzi University, Hangzhou 310000, China
munuan@hdu.edu
[3] College of Computer Science and Technology, National University of Defense Technology, Changsha 410000, China
bkzhao@139.com

Abstract. Software Defined Network (SDN) separates the data layer and control layer of traditional networks, achieving efficient logical centralized control and flexible data forwarding strategy deployment. Compared to traditional networks, the high flexibility and programmability of SDN have made them widely used in various fields such as wide area networks and cloud computing. The successful application of SDN also provides a good reference for the development of multimodal networks in the future; and analyze and predict the endogenous security mechanism of future SDN controller systems. With the continuous development of information technology networks, network service functions have shown diversified and complex characteristics, leading to SDN facing more complex security threats. As the brain and command centre of SDN, controller security protection is important to SDN stable operation. Although mainstream controllers currently have some built-in security protection strategies, due to the lack of targeted defense against special attack methods, they can only serve as a preliminary defense line for SDN security protection. This article starts with the security threats faced by SDN, analyzes the security attacks faced by SDN controllers and the built-in security protection measures of the system, studies the shortcomings of SDN controller systems in facing increasingly diverse security threats, and combines endogenous security to analyze and predict the security protection measures of future SDN controller systems.

Keywords: Software Defined Network · Controller safety protection · Privacy protection · Endogenous security

1 Introduce

The continuous progress of network technology, especially the emergence of new technologies such as industrial internet, autonomous driving, and the Internet

This article is supported by the Key Technology Research and Development Program under Grant No.2022YFB2901204.

of Things, has put forward higher requirements for network diversification [33]. However, the current internet has not fully supported the expansion of network applications when meeting unique service needs, and the constructed network system has a rigid structure and cannot dynamically and flexibly meet diverse network needs through limited resources [82]. SDN is a new type of network architecture that decouples the forwarding surface and control surface of network devices to achieve efficient logical centralized control and flexible data forwarding strategy deployment. The successful application of SDN in various fields such as backbone networks, wide area networks, and cloud computing has provided inspiration for the research of other modal networks. In order to better integrate various modal networks and flexibly deploy network resources, multimodal networks have emerged. Multimodal networks are built on a fully dimensional and definable network support environment, supporting various businesses, services, and other dynamic loading in the form of modalities, achieving the coexistence of multiple network modalities in the same physical environment, thereby meeting diverse network needs [43]. Under the influence of SDN, in order to better meet the research on multimodal networks, many modal networks have shown a trend of software defining everything, such as the emergence of software defined satellite networks (SDSNs) in satellite modal networks [73]. With the continuous development of information technology networks, network service functions have shown diversified and complex characteristics [57], leading to SDN facing more complex security threats. This greatly limits the large-scale promotion of SDN. Therefore, academia and industry are gradually pursuing deeper levels of security protection for SDN.

1.1 Software Defined Network

In the network, hardware dominates, and the control plane and data plane are coupled together in the switch. Packets are forwarded through routing tables, and the forwarding algorithm is also fixed and cannot be changed in the switch. This design approach leads to many drawbacks in the existing network, such as difficulty in deployment management, distributed architecture bottlenecks, difficulty in truly implementing traffic control, non-programmable devices, etc., which means that the network cannot be dynamically adjusted according to requirements. In order to solve the above problems, SDN has emerged. In a 2008 paper [50], Professor Nick McKeown of Stanford University proposed the concept of Openflow for experimental innovation in campus networks. Based on the characteristics of Openflow, it brought programmable features to the network, which led to the concept of SDN. Professor Nick McKeown's goal is to transform a fixed and unchangeable network architecture. SDN aims to separate the network data plane from the control plane and provide an Application Programming Interface (API) to the application layer through a northbound interface, allowing network devices to program directly [5].

The SDN has a three-layer architecture consisting of an application layer, a control layer, and a data layer from top to bottom. The controller of the control layer is at the core of the SDN, which can analyze traffic and update the packet forwarding method of the switch; the data layer is mainly a forwarding device that implements forwarding functions, such as a switch. The application layer implements the network functional applications the business requires, such as network security measures that are basically deployed and applied to the controller. The application program manages devices such as switches in the data layer by calling the controller's northbound interface. SDN uses northbound and southbound APIs for communication between different layers, northbound APIs for communication between application and control layers, and southbound APIs for communication between data and control layers. For multi-controller SDN, communication between controllers is achieved through east-west interfaces [30]. The specific SDN architecture is shown in Fig. 1.

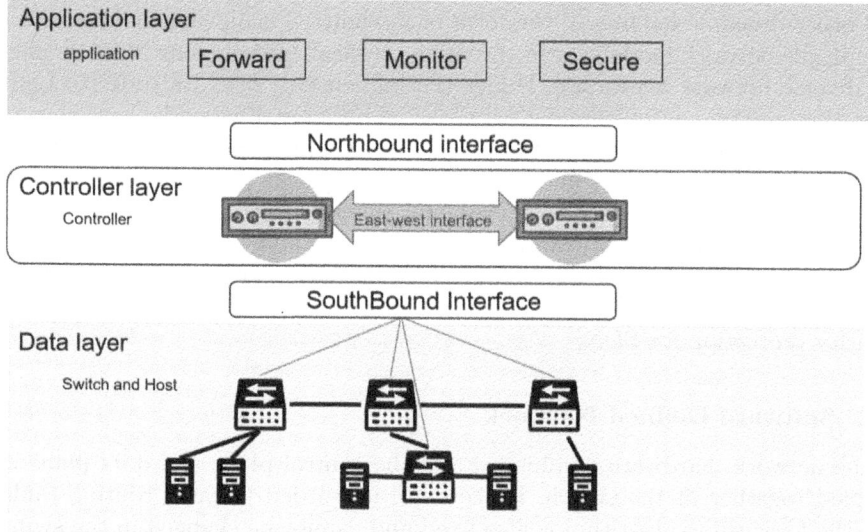

Fig. 1. Software defined network architecture

1.2 Safety Hazards of SDN Controllers

In recent years, the scale of SDN has rapidly expanded, and its functions have become increasingly complex, with diverse and complex network service functions. Although SDN can detect changes or anomalies in the network to a certain extent and provide security protection, the promotion of SDN is gradually limited as security issues become more complex and widespread. As the core of the SDN, the SDN controller is a key target for attackers [6]. At present, there are various types of controllers on the market, and even network attackers are investing in the development of SDN controllers, causing SDN controllers to face very serious vulnerabilities and uncertain security issues [34]. Taking software as an example, OpenSSL released the Heartbleed vulnerability in 2014 [11], which

Table 1. Security vulnerabilities in SDN ontrollers

Vulnerability number	Vulnerable	CVSS3 rating	Details
CVE-2023-30093	ONOS controller	6.1	The arbitrary file upload vulnerability in ONOS 1.9.0 to 2.7.0 allows attackers to execute arbitrary code by uploading crafted YAML files.
CVE-2022-45932	OpenDaylight controller	7.1	OpenDaylight (ODL) before 0.16.5 has a security vulnerability due to the deleteRole function in its RoleStore. java component, which allows attackers to implement SQL injection.

allows network attackers to use OpenSSL vulnerabilities to launch attacks on the network and obtain a large amount of important information. Research [35] has shown that software and hardware vulnerabilities in networks are unavoidable and cannot be thoroughly investigated during installation and design. Therefore, active security protection for SDN controllers is crucial. Table 1 lists some high-risk vulnerabilities in SDN controllers in the Common Vulnerabilities and Exposures (CVE) and China National Vulnerability Database of Information Security (CNNVD) over the years.

In addition to internal vulnerabilities in the controller, more security threats come from external attacks, with the biggest security risks being scanning attacks and Distributed Denial of Service (DDoS) attacks. A scanning attack is the use of reconnaissance by attackers to obtain information about the target, which may include IP addresses, open ports, running services, operating system versions, and network topology. Then, the attacker may use the collected data to prepare for a real attack, such as discovering that the target is running an unpatched application with known vulnerabilities [64]. The SDN controller and switch have a process of message communication, and attackers may use this feature to disguise the switch sending reconnaissance messages to the controller to obtain control information and prepare for future attacks. A variant of Denial of Service (DoS) attacks, called DDoS, is often used to paralyze networks or servers. It involves using a large number of hosts, usually part of a zombie network, all of which work together to paralyze the target network [55]. In SDN, attackers send a large amount of false request information to controllers through different switches, resulting in the paralysis of the entire SDN.

At present, research on SDN security has become relatively mature, and many scholars have analyzed and summarized SDN security threats. Alsmadr et al. [7] conducted extensive research on the security of SDN, discussed the security threats of SDN, and also reviewed a wide range of SDN security controls, such as firewalls and Intrusion Detection Systems (IDS). Finally, they introduced several ways to describe how SDN develops. Ahmad et al. [3]analyzed the security threats faced by the application, control, and data plane of SDN, and then introduced various security methods for network wide security in SDN.

This article emphasizes the current and future security challenges of SDN, as well as the future direction of secure SDN. Scott Hayward et al. [53]conducted a survey on security in SDN and introduced the research community and industry progress in this field. Discussed the challenges of protecting networks from persistent attackers, described the overall approach of security architecture required for SDN, and identified key research directions for providing network security in SDN in the future. Abdou et al. [16] redefined security vulnerabilities in the SDN control plane and provided a comparative analysis framework, comparing the similarities and differences in security between traditional networks and SDN and analyzing the application of security measures in traditional networks on SDN. The above review papers all focus on the comparison between traditional networks and SDN and only focus on passive defense. Compared to previous papers, this article analyzes controller security threats and self-defense measures, and from the perspective of improving the controller's own security protection ability, taking the controller's internal security as the benchmark, provides a reference for the future security protection trend of SDN, and provide a reference for the endogenous security mechanism of multimodal networks in the future. The Sect. 2 of this article will provide a detailed analysis of the security threats faced by the controller. Section 3 analyzes the current defense methods of mainstream controller design and their effectiveness in different security threats. Section 4 introduces two defense methods applicable to SDN architecture: IDS and Moving Target Defense (MTD). Finally, in Sect. 5 , a brief summary of the entire article is provided and suggestions are made for future controller security defense.

2 Controller Security Threats

SDN controllers play a core role in software defined networks. It is the core component of the SDN architecture, responsible for centralized control and management of all network devices in the network, such as switches, routers, and firewalls. If the SDN controller is under security threat, it may lead to many problems and potential risks in the network architecture [48,58]. The controller, as the backbone of the SDN architecture, is an important target for attackers to attack. This section will provide a detailed analysis of the main attacks on the controller.

2.1 Scan Attack

Scanning attack, also known as reconnaissance attack, refers to an attacker scanning the vulnerabilities and weaknesses of the target network or system to find possible attack targets or potential security vulnerabilities, as well as information about the target system, such as running status, memory, and system version [45]. Network scanning attacks are often the first step for attackers to take action, obtaining network architecture information such as switches and controllers, understanding vulnerabilities in the system, and taking further intrusion measures. In addition, attackers sending a large amount of scanning traffic can consume network resources. Therefore, scanning attacks have serious harm to network security [72]. Because the controller communicates with the application and switch through the north and south interfaces, attackers will use this process to launch scanning attacks on the controller. As shown in Fig. 2, attackers can install malicious applications on the controller and obtain information about the controller and network status through the insecure API provided by the controller; attackers can also forge false switches to communicate with the controller to obtain information about the controller.

The possible impacts of scanning attacks include but are not limited to:

1. Information leakage: Attackers may gain access to vulnerabilities and vulnerabilities in the target network or system, in preparation for subsequent attacks.

At present, there have been many studies on the defense against scanning attacks. Ma Yunying [81] uses the distribution characteristics of the destination IP address of the newly created stream on the target host per unit time to identify whether an IP scanning attack has occurred. In this way, attack behavior can be quickly distinguished from normal business flow. This method can quickly

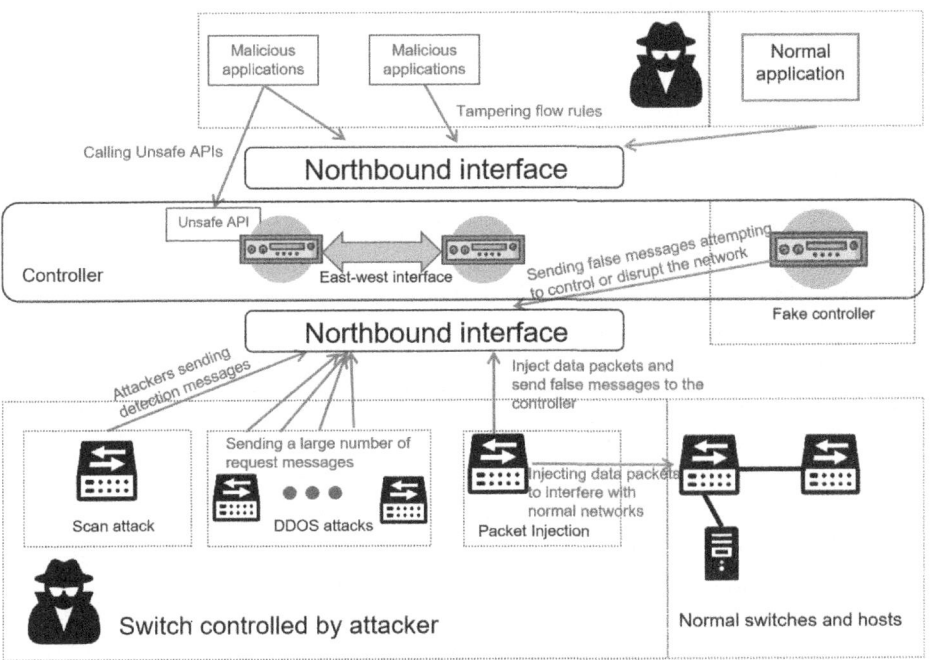

Fig. 2. Attackers launch various attacks on controllers

identify whether the host has been attacked by IP scanning, but its accuracy is not high. Song Hua et al. [66] proposed a new distributed port scanning detection method. The sensor part detects abnormal packets, and the analyzer part divides abnormal packets into different classes. This method can detect many types of scanning attacks, but requires hardware support.

2.2 DDoS Attacks

DDoS attacks are one of the most serious threats facing the Internet and have been a hot topic in network security research in recent years. The principle of this attack is that a large amount of requests or data traffic exceeds the processing capacity of the target system, leading to system resource depletion or network congestion, thereby preventing the service from responding to legitimate user requests properly [55]. DDoS attacks typically involve a large number of attack sources, which may come from controlled switches in SDN. Because the network state is constantly changing in practice, it is difficult for controllers to identify malicious switches, so the security prevention of DDoS attacks in SDN faces enormous challenges. As shown in Fig. 2, attackers can use the controlled switch to send a large number of request packets to the controller, leading to the depletion of controller resources.

The possible impacts of DDoS attacks include but are not limited to:

1. System resource depletion or network congestion: Attackers can use controlled switches to send a large number of request packets to the controller, occupying bandwidth, leading to controller resource depletion and network congestion.
2. Unable to respond properly to legitimate users: DDoS attacks result in system resource depletion and inability to respond to legitimate user requests.

DDoS attack is one of the most difficult attacks to solve in SDN defense, and there is currently a lot of research on defending against DDoS attacks. Machine learning based DDoS detection mechanism and threshold based DDoS detection mechanism are the two most commonly used techniques for detecting DDoS attacks in SDN [15]. Gurusamy et al. [20]proposed a secure traffic management model (SFM) that can dynamically identify and mitigate UDP flooding attacks in multi controller SDNs. Haider et al. [22] proposed a deep convolutional neural network (CNN) integration framework for efficient DDoS attack detection in SDN, which significantly improves the accuracy of DDoS detection compared to manual detection.

2.3 Packet Injection

False data injection attack is a type of attack in deception attacks where attackers inject forged data packets into the target network to alter the normal behavior of the network or achieve specific attack targets [41,84]. As shown in Fig. 2, this attack may lead to serious consequences such as network service interruption, data leakage, and information tampering.

The possible impacts of packet injection attacks include but are not limited to:

1. Data tampering: Attackers can modify the content of data packets, damaging the integrity or reliability of the data.
2. Service interruption: Attackers may send a large number of forged data packets, causing network devices or servers to overload and causing service interruption.
3. Data leakage: Attackers may intercept data packets and steal sensitive information within them.
4. Man in the middle attack: Attackers may become intermediaries in network communication through packet injection attacks, stealing or tampering with data.

Packet injection attacks are attacks that exist in many networks. There has been a lot of research on the prevention of packet injection attacks. Peng Huaye et al. [26] proposed an attack detection mechanism based on analyzing the increment of state measurement values to address the state estimation problem of false data injection attacks in microgrid environments. This method has a high detection success rate in specific systems. Jin Shibo et al. [65] proposed a Bayesian method for detecting false data injection attacks in the Internet of Things, using Bayesian algorithms to determine the type of attack. By comparing feature data with standard data, it is determined whether there is a false data injection attack in the current Internet of Things.

2.4 Flow Rule Tampering

Flow rule tampering is a security threat that acts on SDN controllers, where attackers alter the flow rules in the controller to alter the processing of data packets in the network. The flow rules in SDN are policies defined by the controller to guide the forwarding and processing of data packets in the network.

The possible impacts of stream rule tampering include but are not limited to:

1. Traffic redirection: Attackers can modify flow rules to redirect specific data traffic to other targets, possibly malicious servers or man in the middle attackers.
2. Network isolation failure: The flow rules in SDN are used to achieve network isolation. If an attacker tampers with the rules, it may cause conflicts between networks that should be isolated, reducing network security.
3. Virtual network intrusion: SDN supports the creation of virtual networks, and attackers may enter other virtual or real networks by tampering with flow rules, resulting in data leakage or other security issues.
4. Service quality degradation: By modifying flow rules, attackers can affect the priority and quality of service of network traffic, leading to performance degradation.

Stream rule tampering is a unique attack in SDN, and there are two defense measures against stream rule tampering: 1. Access control restricts physical and remote access to SDN controllers, allowing only authorized administrators to make configuration changes; 2. Encrypted communication uses encrypted communication protocols to protect communication between controllers and network devices, preventing stream rules from being tampered with by intermediaries. Stream rule tampering is a security threat in SDN, and comprehensive measures need to be taken to ensure the security and stable operation of the SDN.

2.5 Fake Controller

A fake controller is a security threat that exists in both single controller and multi controller SDN. In a single controller, it refers to malicious or unauthorized entities impersonating legitimate SDN controllers. In a multi controller, it refers to the presence of downtime or Byzantine points in a cluster. This type of attack may lead to serious consequences such as network operation loss of control, traffic redirection, information leakage, and service interruption. There are two main ways for attackers to launch fake controllers. The first is to impersonate the controller: the attacker creates a fake SDN controller and impersonates a legitimate controller to send commands and flow rules to network devices. Network devices will forward traffic based on these false instructions, leading to abnormal behavior in the network, serving as Byzantine nodes in the cluster to disrupt the cluster [75]. The second type is man in the middle attack: attackers may insert man in the middle devices on the SDN communication link to intercept and tamper with the communication between the controller and network devices, thereby engaging in malicious operations [79].

The potential impacts of fake controller attacks include but are not limited to:

1. Network loss of control: fake controllers may cause communication chaos between network devices by sending false instructions, causing the network to lose control, leading to service interruption or performance degradation.
2. Data traffic tampering: Attackers may tamper with flow rules, redirect network traffic to malicious targets, causing data leakage or denial of service attacks.
3. Security vulnerability exposure: fake controllers may open unauthorized interfaces or services, leading to the exposure of security vulnerabilities in the network.

fake controllers are attacks unique to SDN. There are two defense measures against fake controllers: 1. Encrypting communication. Using encrypted communication protocols to protect communication between controllers and network devices from man in the middle attacks and data tampering. 2. Controller identity verification. Network devices should be able to verify the identity of the controller to ensure connection to legitimate controllers.

2.6 Malicious Applications

In SDN, controllers provide northbound abstraction and APIs, making application development easier. Such abstractions and APIs are not only easy to use, but also powerful because they basically allow applications to do anything they want, and it is indeed necessary to grant such powerful permissions to applications to provide as much network programmability as possible. Such an open design can lead to malicious application threats in the system. Malicious applications refer to applications designed by attackers for malicious purposes, which may silently run in the controller, perform malicious operations on the network, or steal sensitive information. For example, in Fig. 2, attackers use malicious applications for attacks such as Flow Rule Tampering.

The potential threats and impacts of malicious applications include but are not limited to:

1. Network interference: Malicious applications may modify the flow rules in the SDN controller, causing network traffic to be redirected, discarded, or chaotic, resulting in unstable or paralyzed network services.
2. Data leakage: Some malicious applications may be designed to steal sensitive information in the network, such as user data, credentials, etc., posing a threat to user privacy and data security.
3. Information monitoring: Some malicious applications may be used to monitor network communication, steal sensitive information, or monitor user activities.

The malicious application defense includes the following: 1. Application verification. In the SDN controller, only verified and authorized applications are allowed to run. 2. Sandbox environment, running unknown or untrusted applications in an isolated sandbox environment to reduce permissions and limit their impact on the network. 3. Application behavior analysis, detecting abnormal behavior and activity of the application, and timely discovering potential malicious applications.

2.7 Unsafe API

Table 1 indicates that there are unknown vulnerabilities in the controller that may be exploited by attackers. Unsafe APIs refer to potential security vulnerabilities or improperly designed interfaces in the controller that attackers can use to perform unauthorized operations, obtain sensitive information, or pose security risks to the entire SDN. Unsafe APIs provide conditions for malicious applications to obtain more permissions, such as obtaining important information about the network and changing its operation, which may ultimately lead to security issues such as losing important information and network paralysis. As shown in Fig. 2, malicious applications can call insecure APIs to launch attacks.

The potential threats and impacts of insecure APIs include but are not limited to:

1. Unauthorized access: Attackers may use insecure APIs to bypass authentication or access control, thereby performing unauthorized operations such as modifying flow rules, controlling network devices, etc.
2. Data leakage: If the API does not properly validate or process input, attackers may exploit these vulnerabilities to obtain sensitive information or data leakage.
3. Malicious traffic tampering: Attackers may manipulate insecure APIs to alter the flow of traffic in the network, causing packets to be redirected to malicious targets.

The main countermeasures for unsafe APIs include: 1. Safe coding practices. When developing controllers and APIs, follow safe coding practices to ensure that input validation, output filtering, and error handling measures are implemented correctly. 2. Security testing, conducting security testing and code review to identify and fix potential security vulnerabilities. 3. Fix vulnerabilities, promptly patch known security vulnerabilities, and keep controller and API versions updated.

2.8 Brief Summary

The various security threats faced by controllers do not exist separately, but are interconnected. For example, malicious software may lead to scanning attacks, which in turn provide services for DDoS attacks. This article lists the security

Table 2. SDN controller security threats and defense measures

Threat Name	Trigger Subsequent Attacks	Hazard	Preventive Measures
Scan Attack [45,72]	1. DDoS Attacks	1.divulge the information	1.IDS [19,47,66,81] 2.MTD [8,27]
DDoS Attacks [55]	not have	1.System resource depletion or network congestion 2.The service is unable to respond properly to legitimate user requests	1.IDS [15,20,22] 2.Firewall [32,69] 3.Load Balance [44]
Packet Injection [41,84]	1. Fake controller 2. Man-in-the-middle attack	1. Data tampering 2. Service Interruption 3. Data Breach	1.IDS [26,65] 2.Network Encryption and Digital Signature [17,40]
Flow Rule Tampering [1,36]	not have	1. Traffic redirect 2. Network isolation failure 3. Virtual Network Intrusion 4. Reduced service quality	1.Access Control [24,74] 2.Secure Communication [1,67]
Fake controller [75,79]	1.Flow rule Tampering	1. Network out of control 2. Data traffic tampering 3. Security vulnerability exposure	1.Controller Authentication [21,31] 2.Cryptographic Communication [17,40]
Malicious Applications [59,83]	1. Flow Rule Tampering 2. Scan Attack 3. DDoS Attack	1. Network interference 2. Data breach 3. Information monitoring	1.Application Validation [70] 2.Sandbox Mechanism [9,46] 3.Real-Time Monitoring [26,65]
Unsafe API [11,80]	1. Malicious Applications	1. Unauthorized access 2. Data breach 3. Malicious traffic tampering	1.Safety Testing [37,38,60] 2.Bug Fixes [56,76]

threats faced by SDN controllers, potential hazards, and defense measures in Table 2. Among common controller security threats, scanning attacks, DDoS attacks, and packet injection are common attacks in the network. Defense against these attacks can be found in other networks, but how to migrate to SDN and achieve good protection effects is the focus of research. Stream rule tampering, fake controllers, malicious applications, and insecure APIs are unique security threats to SDN. Corresponding prevention measures can be deployed based on the characteristics of controllers to address these security threats, as analyzed in Sect. 3.

3 Analysis of Common Controller Safety Measures

Due to the fact that controllers are an important and vulnerable component of SDN, most controllers on the market have built-in security and prevention functions in their systems [4]. This chapter mainly analyzes the security measures commonly used by ONOS, Ryu, and OpenDaylight controllers to ensure their stable operation.

3.1 ONOS

Identity recognition: The Command Line Interface (CLI), Graphical User Interface (GUI), and REST API of ONOS.

1. ONOS CLI uses public/private key authentication
2. ONOS GUI uses form-based login
3. ONOS REST API uses basic authentication

The REST API is an application programming interface (API or Web API) that follows the REST architecture specification. REST is an English abbreviation for expressive state transfer, created by computer scientist Roy Fielding. In SDN, the REST API is a type of northbound protocol that currently only allows authenticated access and is therefore secure. ONOS also supports role-based differentiated authorization, and deployers can choose to modify the configuration file as needed to change the default role assignment, thereby further limiting or relaxing access requirements. ONOS defines four roles: admin, manager, webconsole, and viewer. The first three roles can execute management and viewing commands, while the viewer can only execute those viewing commands, not those management commands. The identity setting function of ONOS can prevent illegal changes to the controller's configuration to achieve the goal of damaging the SDN. Because the identity setting function has been added, it can effectively prevent attackers from exceeding their authority and prevent flow rules from being tampered with.

Secure Communication: ONOS achieves secure communication by configuring Secure Sockets Layer (SSL) and Transport Layer Security (TLS) with self-signed certificates. TLS/SSL is a specification for encryption channels that utilizes symmetric encryption, asymmetric encryption of public and private keys, and key exchange algorithms to encrypt data for reliable information transmission. In ONOS, it is only necessary to configure correctly on the command line to establish SSL/TLS Openflow secure connections between ONOS controllers and OVS switches, as well as between controllers, ensuring secure communication within the ONOS cluster. Man in the middle attack and packet injection require intercepting information during information transmission or pretending to be a member of the network to send false information. Encrypting the data transmitted in the SDN can effectively defend against man in the middle attack and packet injection attack.

Flow Rule Validation: In ONOS controllers, flow rule validation refers to the process of verifying flow rules (flow table entries) before sending them to network devices. Verify the correctness and legality of flow rules to avoid issues that may lead to network failures or security vulnerabilities. Applications can install flow rules in the ONOS controller through the FlowRuleService API.

Flow rule validation typically includes the following aspects:

1. Syntax Check: Check if the flow rules comply with the Openflow protocol or specific network device Manufacturers specifications. This ensures that the format of the flow rules is correct and there are no spelling errors or other grammar issues.
2. Conflict Check: Check if the new flow rule conflicts with the existing flow rule. If there is a conflict, it may cause traffic to be improperly forwarded or processed.
3. Security Check: Verify if flow rules may cause security vulnerabilities, such as unauthorized access or network segmentation errors.
4. Path Check: Ensure that the flow rule defines a valid network path. This is particularly important as incorrect paths can lead to network loops or invalid traffic forwarding.
5. Efficiency Check: Optimize flow rules to reduce the number of rules and the cost of matching rules, thereby improving network forwarding performance.
6. Fault tolerance Check: Verify the behavior of flow rules in the event of network device failure, ensuring that the network can operate correctly after fault recovery.

The ONOS controller flow rule verification sub module is responsible for performing the above checks. After the flow rules are verified, they will only be distributed to relevant network devices to ensure the stability and security of the network. With the flow rule verification function, the ONOS controller can effectively resist flow rule tampering attacks.

3.2 Ryu

Traffic Monitoring: The network has become the infrastructure of many services or businesses, so maintaining a stable network environment is necessary. But network problems always occur. When there is an abnormality in the network, it is necessary to quickly find the cause in order to recover as quickly as possible.

Only by clearly knowing the status of the network can we identify the errors and the true causes of the network. Therefore, traffic monitoring is essential for the healthy and stable operation of the network. Of course, monitoring network traffic cannot guarantee that no problems will occur. When the system sends DDoS attacks, there will be significant changes in traffic, so traffic monitoring is very helpful for DDoS detection.

Firewall: Firewall technology is a technology that helps computer networks build a relatively isolated protective barrier between their internal and external networks by setting whitelists on software and hardware devices, in order to protect the security of user data and information. In the Ryu controller, network security can be protected by configuring firewalls in both single user and multi user systems, as shown in Table 3, which is an instance of the Ryu firewall whitelist. Ryu controller can use the REST API to set the firewall of the switch, including obtaining and changing the firewall status in the switch, adding, deleting, obtaining switch rules, and obtaining the record status of changing the switch.

Table 3. White list of ICMP protocol firewall for switches

Source Address	Destination Address	Protocol	Action	Rule Id
10.0.0.1/32	10.0.0.2/32	ICMP	pass	1
10.0.0.2/32	10.0.0.1/32	ICMP	pass	2

3.3 OpenDaylight

Secure Communication: There are many separate southbound plugins in the OpenDaylight controller that provide a mechanism to protect its security and communicate with network devices. For example, the Openflow plugin supports TLS connections with bidirectional authentication and the NETCONF plugin supports SSH connections. Security plugins provide a way to form secure remote connections for supported devices to ensure secure communication between switches and controllers, as well as between controllers. Therefore, when deploying OpenDaylight, it is important to understand the relevant plugin mechanisms and deploy them. Like ONOS, OpenDaylight encrypts data transmitted in SDN to effectively defend against man in the middle attacks and packet injection attacks.

Authentication: OpenDaylight uses AAA to protect the security of the controller, representing authentication, authorization, and accounting. All three services contribute to the secure deployment of OpenDaylight. The vast majority of OpenDaylight's northbound APIs are protected by AAA by default when installing this feature. Malicious applications require the use of APIs provided by the controller in order to function effectively, so OpenDaylight's AAA mechanism can not only prevent unsafe APIs but also effectively prevent malicious applications.

The background of the birth of different controllers varies, and the applicable network environment also varies. Therefore, the security functions they provide also vary. Table 4 lists the security measures and threats that different controllers come with.

4 Future Development Trends of SDN Controller Security Defense

Due to the controller's limited understanding of specific attacks and the constantly changing nature of the attacks, what the controller can only provide is relatively broad and less targeted security protection functions such as identity verification. From Table 4, it can be seen that the built-in security protection function of the controller can play a role in some threats. However, for flexible attacks such as scanning attacks and DDoS attacks, the built-in security protection function of the controller has not achieved good results, resulting in serious challenges for the security of SDN controllers [54]. The uncertain security threats caused by vulnerabilities and backdoors are the most serious, difficult, and incompletely solvable problems in cyberspace [71]. Traditional defense methods such as firewalls and identity authentication face a lack of focus in the face of explosive network development and the combination of different modal networks. Therefore, the most suitable defense method should be studied from the perspective of endogenous security, which relies on the network's own construction factors to cope with the rapid development and endless attacks of the network. In SDN, controllers have a global view and can configure network monitoring resources, so implementing IDS in SDN is of great research value. The SDN controller, as the brain of the network, can change the network configuration and provide a prerequisite for the implementation of MTD. Therefore, compared to traditional networks, SDN is more suitable for the application of MTD. Considering the characteristics of SDN architecture, this chapter introduces two commonly used methods for addressing controller security threats: IDS and MTD.

Table 4. Built-in security measures and threats that can be addressed

	Authentication	Secure Communication	Flow Rule Validation	Traffic Monitoring	Firewall
ONOS	1. Flow rule tampering 2. Malicious applications	1. Man-in-the-middle 2. Packet Injection	Flow rule tampering	Without this feature	Without this feature
Ryu	Without this feature	Without this feature	Without this feature	DDoS Attack	Packet Injection
OpenDaylight	1. Flow rule tampering 2. Malicious applications	1. Man-in-the-middle 2. Packet Injection	Without this feature	Without this feature	Without this feature

4.1 IDS

IDS, as a security measure, aims to detect malicious activities and security vulnerabilities in computer systems, networks, or applications. It detects potential intrusion behavior by monitoring the activity of computer systems, networks, or applications, and issues alerts to notify administrators to take appropriate measures. In traditional networks, it can be divided into network-based detection and host-based detection based on the location of detection. Network-based detection aims to monitor network traffic to detect potential intrusion behavior and security vulnerabilities. Host-based IDS aim to detect malicious activities and security vulnerabilities on a single computer or server. The controller can obtain the entire network status, and the network administrator can install applications on the controller according to requirements. This two flexible characteristics of SDN provide superior conditions for the implementation of IDS compared to traditional networks, and IDS can effectively respond to scanning attacks and DDoS attacks. Therefore, IDS will still be a hot research direction in SDN security protection.

The SDN controller has a global view, so IDS can be performed through the controller. Braga et al. [10] proposed a lightweight method for detecting DDoS attacks based on traffic features, which has a very low cost of extracting such information compared to traditional methods. SM Mousavi et al. [51]proposed using central control of SDN for attack detection and introduced an effective and lightweight solution for resource utilization. And provides a solution for detecting DDoS attacks based on the entropy change of the target IP address. Shihabur et al. [13] focused on the trade-off between monitoring accuracy, timeliness, and network overhead. A monitoring framework for PayLess-SDN has been proposed, which provides a flexible RESTful API for collecting traffic statistics at different aggregation levels. It uses adaptive statistical collection algorithms to provide highly accurate information in real-time without significant network overhead. Although they [10,13,51] deployed IDS in control, the accuracy of detection was not high. Artificial intelligence has made significant progress in recent years and plays an important role in detection. Therefore, many scholars have applied artificial intelligence to IDS in SDN. Yang [77] has designed an IDS model based on convolutional neural networks to solve the IDS problems faced in SDN environments. Compared to methods that did not use artificial intelligence, Yang's designed model significantly improves the detection accuracy of DDoS attacks. The distributed nature of networks poses two major challenges for detecting abnormal network traffic: uneven data distribution leads to model training being unable to optimize, and distributed training of global models is not suitable for local network abnormal traffic detection. Chen Hexiong et al. [12]proposed a collaborative anomaly traffic detection technology based on federated learning, combining the topology and traffic characteristics of SDN and extracting traffic features by calculating information entropy. Compared with the local independent training algorithm, the method proposed by Chen has improved the setting accuracy by 31.69%.

The switch communicates with the controller through a southbound interface, but due to the switch sending a large amount of received traffic information to the controller, the southbound interface has a significant delay and communication overhead. This makes it easy for the efficiency of attack detection methods through controllers to reach bottlenecks. So choosing a data plane to participate in traffic detection to reduce the communication cost of the southbound interface has also become a key research object in the academic community. Han et al. [23]proposed a new defense strategy offloading mechanism that can dynamically deploy defense applications on controllers and switches, enabling fast attack response and accurate botnet localization. This method has high detection accuracy and real-time DDoS attack response, and reduces communication overhead on the SDN southbound interface. When DDoS attacks occur, the entropy value in the network will undergo significant changes, so switch pre detection can be performed based on entropy value. Mousavi et al. [52] proposed an entropy based early detection method for DDoS. It is a lightweight detection method that counteracts DDoS attacks by calculating the entropy of the destination IP address in the SDN controller, but this method can only support a single controller architecture. Sahoo et al. [61] proposed a measure based on generalized entropy (GE) to detect low-rate DDoS attacks on the control layer, utilizing the traffic based characteristics of SDN. Compared with Shannon entropy and other statistical distance measures, the GE based detection mechanism improves detection accuracy. For scanning attacks, attackers will send a large number of scanning attack packets to the controller, so scanning attacks can be identified based on the frequency of sending scanning packets. However, there are bottlenecks in the accuracy of recognition through entropy and scanning frequency, making it difficult to achieve higher recognition rates. The detection ability of IDS is also affected by packet capture mechanisms. In practice, SDN have a large amount of traffic, which poses a challenge in how to capture packets in large amounts of traffic. Schaelicke et al. [62] research suggests that the detection ability of IDS decreases with packet loss. Hu et al. [25] designed a packet capturing mechanism that can use DPDK in high-speed networks to capture important data packets.

IDS consumes a large amount of network resources and controller computing resources, so ensuring a balance between detection effectiveness and resource utilization is a long-term research direction. Kim et al. [39] proposed an adaptive traffic detection and MTD adversarial framework based on deep reinforcement learning (DRL), which can allocate detection capabilities to more vulnerable devices or suspicious traffic to achieve optimal system utilization. Figure 3 shows a schematic diagram of IDS based on switch pre detection and controller detection.

4.2 MTD

There is information asymmetry between the attacker and the defender during the confrontation, and the attacker must be familiar with the defender when launching the attack. However, the defender's defense methods based on prior

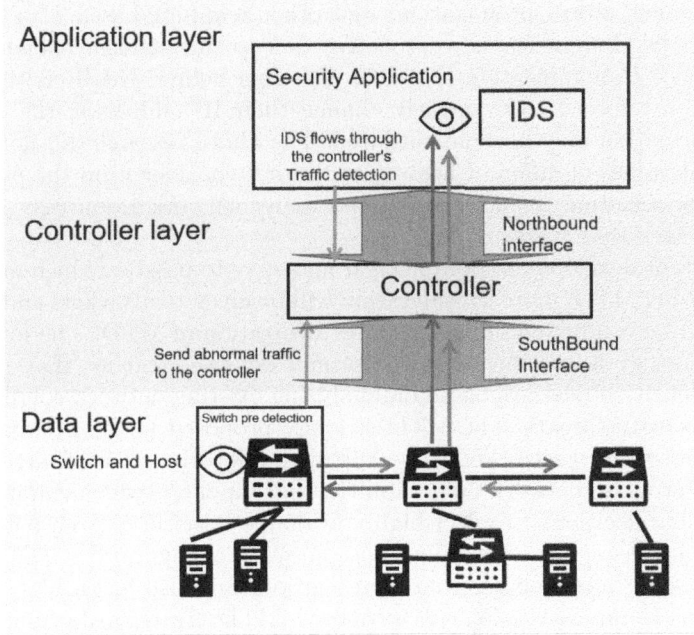

Fig. 3. Schematic diagram of SDN intrusion detection

knowledge are difficult to list all possible attack methods and discover all potential resource vulnerabilities. The static characteristics of network structure provide conditions for attackers to implement intrusions, and the homogeneity of network elements provides a survival space for attackers to expand the scope of destruction. MTD alleviates the asymmetry of attack and defense information by changing the attack surface, increasing the distress of attackers [28,42]. Therefore, MTD can effectively prevent many types of attacks, such as DDoS attacks and scanning attacks. The SDN controller has a global view, and as the brain of the network, the ability to change the network configuration provides a prerequisite for the implementation of MTD. However, how to achieve the best defense effect of MTD to prevent unknown attacks is another direction for the development of SDN security.

In the initial research on MTD as an active defense, scholars did not consider attackers, but instead changed the network's attack surface according to a fixed frequency. The static configuration provides significant advantages for opponents to discover network targets and launch attacks, so Jafarian et al. [27] developed an MTD architecture using OpenFlow, which transparently changes host IP addresses with high unpredictability and high speed. The proposed technology is called OpenFlow Random Host Mutation (OF-RHM), where the OpenFlow controller frequently assigns random virtual IPs to each host, while the true IP remains unchanged. The results indicate that OF-RHM can effectively resist stealth scanning, worm propagation, and other scanning-based attacks. Antonatos et al. [8] designed a new proactive defense mechanism called Network Address Space Randomization (NASR). The idea behind NASR is that if network nodes are forced to frequently change their IP addresses, the attacker's hit list information may become outdated. The above research did not consider the different values of different hosts, and did not consider from the perspective of attackers, resulting in MTD using too many network resources, and many transformations were ineffective.

The current development of MTD is from active to passive, which means that when deploying MTD, more consideration will be given to attackers and strategic analysis will be conducted on the resource utilization of MTD. The formulation of MTD strategy is to generate the optimal defense strategy that meets the expected security objectives based on analyzing existing network conditions and potential security threats. The MTD strategy proposed in the literature [8,27] can cause resource waste. Yoon et al. [78] developed a layered attack graph model that provides network vulnerabilities and topology, which can be used for MTD shuffling decisions to select highly available hosts in a given network and determine the frequency of shuffling host network configuration. The proposed method not only has higher resource utilization, but also has a higher defense success rate. Compared to Yoon's proposed MTD strategy, Javadpour et al. [29] proposed a cost effective edge based MTD method (SCEMA) based on SDN, which reduces DDoS attacks at a lower cost by transforming a set of optimized hosts for critical servers with the highest number of connections. The results indicate that it is targeted against specific attacks. SCEMA has better complexity and performance than Yoon.

The difficulty in implementing MTD on traditional networks lies in how to change the attack surface, so MTD has not played a good role in traditional networks. Compared to the difficult to change characteristics of traditional networks, in SDN networks, network administrators can change the network's attack surface according to requirements through the application layer, providing more options for implementing MTD. [27,49] achieve MTD by transforming or randomizing IP addresses; [68] achieves MTD by randomizing headers; [2] achieve MTD by changing topology changes. The commonality of [2,27,49,68] is achieved through SDN controllers, which shows that SDN technology has been effectively deployed with various MTD technologies. Figure 4 is a schematic diagram of MTD's security protection on SDN.

As shown in Table 5, IDS and MTD both have advantages and disadvantages in different research directions. The current research direction of IDS is divided into switch based IDS and controller based IDS. Based on switch IDS, there is no need to send packets to the controller for processing, thus saving communication resources from the switch to the controller. Due to the low computing power of the switch, the accuracy of detection is poor. Due to the high computing power of the controller, the detection accuracy of IDS based on the controller is relatively high. Active MTD does not take into account attackers and generally performs random transformations on the network attack surface, so the

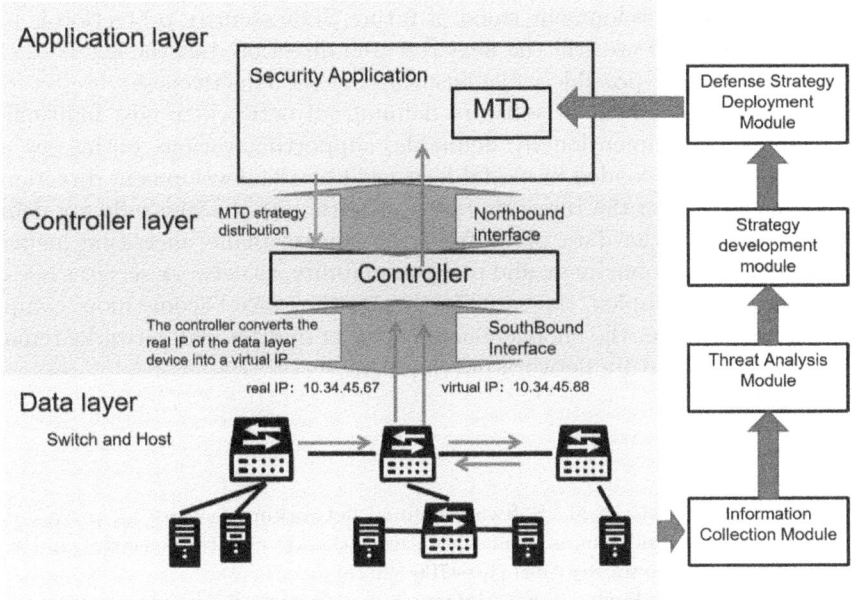

Fig. 4. MTD Defense Deployment in SDN

disadvantage of this method is that it wastes resources. Passive MTD takes into account the attacker's thoughts and focuses on defense against high-risk devices, so this method can achieve good defense effects while also maximizing resource savings.

Table 5. Different research directions of IDS and MTD in ensuring controller security

Means Of Defense	Research Direction	Advantage	Disadvantage
IDS	Switch Based [10,12,13,51,77]	Save Resources	Low Accuracy
	Controller Based [23,25,39,52,61,62]	High Accuracy	Consuming Resources
MTD	Active [8,27]	Excellent effect	Consuming Resources
	Passive [14,18,29,63,78]	Save Resources	Algorithm Complexity

5 Summary

This article first introduces the current security threats faced by SDN controllers and analyzes their impact; Then, based on the analysis of the security protection effectiveness of the existing SDN controller's built-in security defense measures, it is concluded that the current mainstream SDN controller's built-in security protection measures are insufficient to protect against special means of attack;

On this basis, the development trend of future SDN security protection is analyzed, IDS and MTD are still the next research direction, but the key is to save resources as much as possible while ensuring defense effectiveness.

With the development of software defining all network trends, multimodal networks, as fully dimensionally definable, supporting various businesses and services dynamically loaded in modal form, will be the development direction of the next generation of the Internet. However, although the way software defines networks separates the data plane from the control plane, increasing network systems' flexibility, robustness, and programmability, as network services become more diverse and complex, the security threats they face become more complex and varied. Therefore, the endogenous security of multimodal networks remains a key direction for future network development.

References

1. Abdelrahman, A.M., et al.: Software-defined networking security for private data center networks and clouds: vulnerabilities, attacks, countermeasures, and solutions. Int. J. Commun Syst **34**(4), e4706 (2021)
2. Achleitner, S., La Porta, T.F., McDaniel, P., Sugrim, S., Krishnamurthy, S.V., Chadha, R.: Deceiving network reconnaissance using SDN-based virtual topologies. IEEE Trans. Netw. Serv. Manage. **14**(4), 1098–1112 (2017)
3. Ahmad, I., Namal, S., Gurtov, A., Ylianttila, M.: Security in software defined networks: a survey. Commun. Surv. Tutorials **17**(4), 2317–2346 (2015)
4. Ahmad, S., Mir, A.H.: Scalability, consistency, reliability and security in SDN controllers: a survey of diverse SDN controllers. J. Netw. Syst. Manage. **29**, 1–59 (2021)
5. Al-Adaileh, M.A., Anbar, M., Chong, Y.W., Al-Ani, A.: Proposed statistical-based approach for detecting distribute denial of service against the controller of software defined network (SADDCS). In: MATEC Web of Conferences, vol. 218, p. 02012. EDP Sciences (2018)
6. Aladaileh, M.A., Anbar, M., Hasbullah, I.H., Chong, Y.W., Sanjalawe, Y.K.: Detection techniques of distributed denial of service attacks on software-defined networking controller-a review. IEEE Access **8**, 143985–143995 (2020)
7. Alsmadr, I., Xu, D.: Security in software defined networks: a survey. Comput. Secur. **53**, 79–108 (2015)
8. Antonatos, S., Akritidis, P., Markatos, E.P., Anagnostakis, K.G.: Defending against hitlist worms using network address space randomization. In: Proceedings of the 2005 ACM Workshop on Rapid Malcode, pp. 30–40 (2005)
9. Bin, L.: Design and Implementation of a Software Behavior Analysis System Based on Android Sandbox. Ph.D. thesis, Beijing University Of Posts and Telecommunications (2013)
10. Braga, R., Mota, E., Passito, A.: Lightweight DDoS flooding attack detection using NOX/OpenFlow. In: IEEE Local Computer Network Conference, pp. 408–415. IEEE (2010)
11. Carvalho, M., Demott, J., Ford, R., Wheeler, D.A.: Heartbleed 101. IEEE Secur. Priv. **12**(4), 63–67 (2014)
12. Hexiong, C., et al.: A federated learning based collaborative detection method for abnormal traffic in SDN networks. Comput. Eng. **49**(3), 9 (2023)

13. Chowdhury, S.R., Bari, M.F., Ahmed, R., Boutaba, R.: PayLess: a low cost network monitoring framework for software defined networks. In: 2014 IEEE Network Operations and Management Symposium (NOMS), pp. 1–9. IEEE (2014)
14. Clark, A., Sun, K., Bushnell, L., Poovendran, R.: A game-theoretic approach to IP address randomization in decoy-based cyber defense. In: Khouzani, M.H.R., Panaousis, E., Theodorakopoulos, G. (eds.) GameSec 2015. LNCS, vol. 9406, pp. 3–21. Springer, Cham (2015). https://doi.org/10.1007/978-3-319-25594-1_1
15. Cui, Y., Qian, Q., Guo, C., Shen, G., Yan, L.: Towards DDoS detection mechanisms in software-defined networking. J. Netw. Comput. Appl. **190**(2), 103156 (2021)
16. Deb, R., Roy, S.: A comprehensive survey of vulnerability and information security in SDN. Comput. Netw. **206**, 108802 (2022)
17. Dongfeng, C., Xiaoxin, H.: Blockchain based asymmetric encryption and decryption mechanism for data. Netw. Secur. Technol. Appl. (10), 2 (2022)
18. Feng, X., Zheng, Z., Mohapatra, P., Cansever, D.: A stackelberg game and Markov modeling of moving target defense. In: Decision and Game Theory for Security: 8th International Conference, GameSec 2017, Vienna, Austria, October 23-25, 2017, Proceedings, pp. 315–335. Springer (2017). https://doi.org/10.1007/978-3-319-68711-7_17
19. Gang, R., Yu, Z.: Scanning and information collection attacks - security threats still to be faced with in the next generation internet. China Internet (9), 2 (2004)
20. Gurusamy, U., K, H., MSK, M.: Detection and mitigation of UDP flooding attack in a multi controller software defined network using secure flow management model. Concurrency Comput. Pract. Exper. **31**(20), e5326 (2019)
21. Haggag, M., Tantawy, M.M., El-Soudani, M.M.: Token-based authentication for Hadoop platform. Ain Shams Eng. J. **14**(4), 101921 (2023)
22. Haider, S., et al.: A deep CNN ensemble framework for efficient DDoS attack detection in software defined networks. IEEE Access **8**, 53972–53983 (2020)
23. Han, B., Yang, X., Sun, Z., Huang, J., Su, J.: OverWatch: a cross-plane DDoS attack defense framework with collaborative intelligence in SDN. Secur. Commun. Netw. **2018**, 1–15 (2018)
24. Han, Y., Li Junni, L.W.: Xuandonghai: Blockchain based access control scheme for energy data sharing. J. Inf. Secur. Res. **9**(3), 8 (2023)
25. Hu, Q., Yu, S.Y., Asghar, M.R.: Analysing performance issues of open-source intrusion detection systems in high-speed networks. J. Inf. Secur. Appl. **51**, 102426 (2020)
26. Huaye, P., Peng Chen, S.H., Mingjin, Y.: Incremental detection mechanism for microgrids under false data injection attacks. Inf. Control **48**(5), 6 (2019)
27. Jafarian, J.H., Al-Shaer, E., Duan, Q.: OpenFlow random host mutation: transparent moving target defense using software defined networking. In: Proceedings of the First Workshop on Hot Topics in Software Defined Networks, pp. 127–132 (2012)
28. Jalowski, Ł, Zmuda, M., Rawski, M.: A survey on moving target defense for networks: a practical view. Electronics **11**(18), 2886 (2022)
29. Javadpour, A., Ja'fari, F., Taleb, T., Shojafar, M., Yang, B.: SCEMA: an SDN-oriented cost-effective edge-based MTD approach. IEEE Trans. Inf. Forensics Secur. **18**, 667–682 (2022)
30. Jefia, A., Popoola, S.I., Atayero, A.A.: Software-defined networking: current trends, challenges, and future directions, pp. 1677–1685 (2018)
31. Jeong, P.S., Cho, Y.H.: Multiple method authentication system using embedded device. J. Korea Convergence Soc. **10**(7), 7–14 (2019)

32. Jianguo, Z., Huan, C.: Research on DDoS hardware firewall technology. Netinfo Security (12), 2 (2010)
33. Jiangxing, W.: New network architecture. J. Commun. (05), 181 (2014)
34. Jiangxing, W.: Constructing national information cyberspace endogenous security through pseudomorphic defense technology. Inf. Commun. Technol. **13**(6), 3 (2019)
35. Jiangxing, W.: The paradigm of endogenous security development in cyberspace. Chin. Sci. Inf. Sci. **52**(2), 189–204 (2022)
36. Jing, J., Zhi, X.: The principle and prevention of SSL man in the middle attack. Inf. Secur. Commun. Priv. (4), 3 (2007)
37. Jinxia, A., Wang Guoqing, L.S., Jihong, Z.: A dynamic evaluation method for software testing based on multidimensional coverage. J. Software (9), 13 (2010)
38. Jiong, Y., Ji, W., Huowang, C.: Overview of model based software testing. Comput. Sci. **31**(2), 4 (2004)
39. Kim, S., et al.: DIVERGENCE: deep reinforcement learning-based adaptive traffic inspection and moving target defense countermeasure framework. IEEE Trans. Netw. Serv. Manage. **19**(4), 4834–4846 (2022)
40. Kittur, A.S., Pais, A.R.: Batch verification of digital signatures: approaches and challenges. J. Inf. Secur. Appl. **37**, 15–27 (2017)
41. Lanzhi, F.: Design and Detection of False Data Injection Attacks in Networked Control Systems. Ph.D. thesis, North China University of Technology (2021)
42. Lei, C., Zhang, H.Q., Tan, J.L., Zhang, Y.C., Liu, X.H.: Moving target defense techniques: a survey. Secur. Commun. Netw. **2018** (2018)
43. Li, J., Hu, Y., Cui, P., Tian, L., Dong, Y.: Research on incremental deployment mechanism of network modality for multimodal network environment. J. Commun. **35**(08), 184–197 (2014)
44. Lijuan, T., Yongping, Z., Lili, Z.: A comprehensive and proactive defense scheme against DDoS attacks. Microcomput. Inf. (18), 3 (2007)
45. Lin, M.: Implementation of port scan detection technology. China CIO News (3), 2 (2013)
46. Long, C., Xiaohu, Y.: Implementation of sandbox module in Linux system kernel. J. Comput. Appl. **24**(1), 3 (2004)
47. Longye, W.: Roger: Security detection methods for internet port scanning attacks. Inf. Secur. Technol. (2), 3 (2016)
48. Luo, S., Wu, J., Li, J., Pei, B.: A defense mechanism for distributed denial of service attack in software-defined networks. In: 2015 Ninth International Conference on Frontier of Computer Science and Technology, pp. 325–329. IEEE (2015). https://doi.org/10.1109/FCST.2015.11
49. MacFarland, D.C., Shue, C.A.: The SDN shuffle: creating a moving-target defense using host-based software-defined networking. In: Proceedings of the Second ACM Workshop on Moving Target Defense, pp. 37–41 (2015)
50. McKeown, N., et al.: OpenFlow: enabling innovation in campus networks. ACM SIGCOMM Comput. Commun. Rev. **38**(2), 69–74 (2008)
51. Mousavi, S.M., St-Hilaire, M.: Early detection of DDoS attacks against SDN controllers. In: 2015 International Conference on Computing, Networking and Communications (ICNC), pp. 77–81. IEEE (2015)
52. Mousavi, S.M., St-Hilaire, M.: Early detection of DDoS attacks against software defined network controllers. J. Netw. Syst. Manage. **26**, 573–591 (2018)
53. Natarajan, S., Scott-Hayward, S., Sezer, S.: A survey of security in software defined networks. Commun. Surv. Tutorials **18**(1), 623–654 (2016)

54. Nisar, K., Welch, I., Hassan, R., Sodhro, A.H., Pirbhulal, S.: A survey on the architecture, application, and security of software defined networking. Internet Things **12**(5), 100289 (2020)
55. Peng, L.: Research on the principles and defense mechanisms of DDoS attacks. Commun. Technol.(4), 3 (2010)
56. Peng, Z., Yanjun, W., Chen, Z.: An automatic identification method for Linux security vulnerability repair patches. J. Comput. Res. Dev. **59**(1), 12 (2022)
57. Quan, R.: Research on the Construction and Key Technologies of Endogenous Security Control for Software Defined Networks. Ph.D. thesis, Information Engineering University (2023)
58. Revathi, S., Geetha, A., et al.: A survey of applications and security issues in software defined networking. Int. J. Comput. Network Inf. Secur. **9**(3), 21 (2017)
59. Rihuang, Y., Xun, L., Haiyang, G.: Research on malicious application detection technology based on network traffic detection. Electron. Qual. **06**(7), 68–72 (2023)
60. Ruifang, M., Huiran, W.: Research on computer software testing methods. J. Chin. Comput. Syst. **24**(12), 4 (2003)
61. Sahoo, K.S., Puthal, D., Tiwary, M., Rodrigues, J.J., Sahoo, B., Dash, R.: An early detection of low rate DDoS attack to SDN based data center networks using information distance metrics. Futur. Gener. Comput. Syst. **89**, 685–697 (2018)
62. Schaelicke, L., Freeland, J.C.: Characterizing sources and remedies for packet loss in network intrusion detection systems. In: IEEE International. 2005 Proceedings of the IEEE Workload Characterization Symposium, pp. 188–196. IEEE (2005)
63. Sengupta, S., Chowdhary, A., Huang, D., Kambhampati, S.: Moving target defense for the placement of intrusion detection systems in the cloud. In: Bushnell, L., Poovendran, R., Başar, T. (eds.) GameSec 2018. LNCS, vol. 11199, pp. 326–345. Springer, Cham (2018). https://doi.org/10.1007/978-3-030-01554-1_19
64. Sengupta, S., Chowdhary, A., Sabur, A., Alshamrani, A., Huang, D., Kambhampati, S.: A survey of moving target defenses for network security. IEEE Commun. Surv. Tutorials **22**(3), 1909–1941 (2020)
65. Shibo, J., Li, Z.: Bayesian based detection of false data injection attacks in the internet of things. Comput. Simul. **39**(11), 406–410 (2022)
66. Song Hua, L.P., Yiqi, D.: A new distributed port scan detection method. Comput. Eng. Appl. **39**(8), 4 (2003)
67. Ullah, I., Khan, M.A., Abdullah, A.M., Noor, F., Innab, N., Chen, C.M.: Enabling secure communication in wireless body area networks with heterogeneous authentication scheme. Sensors **23**(3), 1121 (2023)
68. Wang, Y., Chen, Q., Yi, J., Guo, J.: U-TRI: unlinkability through random identifier for SDN network. In: Proceedings of the 2017 Workshop on Moving Target Defense, pp. 3–15 (2017)
69. Wei, W.: Design of DDoS firewall based on NDIS middle layer driver. Exp. Sci. Technol. **13**(2), 4 (2015)
70. Wen, X., Zhang, W., Y.Y., Juan, H.: Research on BPEL application verification model. Comput. Sci. **36**(4), 3 (2009)
71. Wu, J.: Cyberspace endogenous safety and security. Sci. China (8), 7 (2022)
72. Xingxuan, L., Li, H.: Research on scan attack detection model based on PCA-DNN in SDN environment. J. Tianjin Univ. Technol. (038-001) (2022)
73. Xu, F., Zhou, X., Zhao, J., Wu, F., Lin, Y., Xia, Y.: Concept and development of software defined satellite technology. J. Beijing Univ. Aeronaut. Astronaut. **49**(07), 1543–1552 (2023)
74. Yakoob, S., Reddy, V.K.: Efficient identity-based multi-cloud security access control in distributed environments. Int. J. e-Collaboration (IJeC) **19**(3), 1–13 (2022)

75. Yin Tuokai, Y.W., Zhi, C.: Cognitive user classification for byzantine attacks. Comput. Technol. Dev. **33**(4), 102–107 (2023)
76. Yingping, H.: Analysis of computer security vulnerability detection and vulnerability repair technology solutions. China Manage. Inform. (18), 2 (2017)
77. Yintan, Y.: Research on SDN Intrusion Detection Technology Based on Convolutional Neural Networks. Ph.D. thesis, Xidian University (2019)
78. Yoon, S., Cho, J.H., Kim, D.S., Moore, T.J., Free-Nelson, F., Lim, H.: Attack graph-based moving target defense in software-defined networks. IEEE Trans. Netw. Serv. Manage. **17**(3), 1653–1668 (2020)
79. Yuan, W.: Design and Implementation of Penetration Testing System Based on Minimizing Attack Graph. Ph.D. thesis, Second Research Institute of the China Aerospace Science and Industry Group 2 (2014)
80. Yuchen, W.: System vulnerability management and common attack methods. Comput. Eng. Appl.(03), 62–64+92 (2001)
81. Yunying, M.: IP scan attack identification technology based on address distribution features. Inf. Commun. (10), 3 (2017)
82. Yuxiang, H., Yi Peng, S.P., Jiangxing, W.: Research on a fully dimensional and definable multimodal intelligent network system. J. Commun. **40**(8), 12 (2019)
83. Zhida, S., Yuefei, Z., Long, L.: Android malicious application detection based on deep learning. J. Comput. Appl. **37**(6), 7 (2017)
84. Zhuowei, W.: Analysis of man in the middle attack against weak encryption algorithms in SSH services. J. Fujian Comput. **38**(12), 49–52 (2022)

Author Index

C
Chang, Harry 146
Chen, Cong 61
Chen, Jia 112
Chen, Jing 1
Chen, Jue 99
Chen, Kai 112
Chen, Xu 33
Cong, Ligang 1
Cui, Meng 99

D
Deng, Xiaoheng 84
Di, Xiaoqiang 1
Dong, Qinghe 74

G
Guo, Kuo 112

H
He, Qian 74, 167
Huang, Xu 112

L
Li, Anyi 128
Li, Danqi 146
Li, Jingyi 33
Li, Jinqing 1
Li, Yangyang 182
Li, Youhuizi 182
Liang, Xiangbin 128
Liao, Bingjie 74
Liao, Chenxi 112
Liu, Jingjing 112
Liu, Peng 74
Liu, Shang 112
Liu, Xiaolong 112
Liu, Ximeng 61

Liu, Xuhui 84
Liu, Yanbo 167
Liu, Yanhua 61
Liu, Zhaoyang 17
Lv, Yusheng 112

O
Ouyang, Bei 33

P
Pan, Qi 167
Peng, Jiajun 167

Q
Qi, Hui 1
Qian, Dongsheng 112
Qiu, Kun 146

S
Shen, Jiajie 48

T
Tian, Ningbo 146

W
Wang, Rui 182
Wang, Tianshu 128
Wu, Bochun 48

X
Xiang, Wang 48
Xue, Meiting 182
Xue, Ru 17

Y
Yin, Yuyu 182
Yu, Xiahui 146

Z
Zeng, Fanhao 61
Zeng, Zengri 84
Zhang, Kai 48

Zhang, Qiu 61
Zhao, Baokang 84, 128, 182
Zhao, Jin 146
Zhao, Zeyu 48

SPRINGER NATURE

GPSR Compliance

The European Union's (EU) General Product Safety Regulation (GPSR) is a set of rules that requires consumer products to be safe and our obligations to ensure this.

If you have any concerns about our products, you can contact us on ProductSafety@springernature.com

In case Publisher is established outside the EU, the EU authorized representative is:

Springer Nature Customer Service Center GmbH
Europaplatz 3
69115 Heidelberg, Germany

The manufacturer's authorised representative in the EU is Springer Nature Customer Service Centre GmbH, Europaplatz 3, 69115 Heidelberg, Germany. If you have any concerns regarding our products, please contact ProductSafety@springernature.com

Printed and bound by CPI Group (UK) Ltd, Croydon, CR0 4YY

25/03/2026

02078185-0009